TheSky™
Workbook

Thomas Jordan
Ball State University

Scott Peters

BROOKS/COLE

THOMSON LEARNING

Australia • Canada • Mexico • Singapore • Spain • United Kingdom • United States

BROOKS/COLE

THOMSON LEARNING

Sponsoring Editor: Keith Dodson
Assistant Editor: Samuel Subity
Marketing Assistant: Maureen Griffin
Editorial Assistant: Faith Riley
Production Coordinator: Stephanie Andersen
Manuscript Editor: Linda Purrington

Permissions Editor: Sue Ewing
Cover Design: Denise Davidson
Cover Photo: Corbis
Print Buyer: Nancy Panziera
Printing and Binding: Transcontinental Printing

For more information about this or any other Brooks/Cole products, contact:
BROOKS/COLE
511 Forest Lodge Road
Pacific Grove, CA 93950 USA
www.brookscole.com
1-800-423-0563 (Thomson Learning Academic Resource Center)

For permission to use material from this work, contact us by
www.thomsonrights.com
fax: 1-800-730-2215
phone: 1-800-730-2214

Printed in Canada

10   9   8   7   6   5   4

ISBN: 0-534-39072-2

Preface

The workbook you are about to use is a non-traditional workbook. By this, we mean that it is a stand-alone instruction manual. Many of the concepts presented in this workbook make using *TheSky* easy. It has been designed for use in an introductory observational astronomy course or for the causal observer.

Many traditional workbooks give brief and sometimes incomplete information about astronomical topics. They tend to dwell more on how to use the software and less so on the concepts. This workbook employs an in-depth approach in discussing many astronomical topics. It gives detailed explanations of difficult concepts. It is written with the assumption that most students who will use this workbook have little or no knowledge of the concepts used in observational astronomy.

We, the authors, have given special attention to details in how to fully use the software, with many examples, exercises, and multimedia sky windows. All have been integrated into the chapters throughout the workbook. At the end of each chapter review exercises and questions help students test their understanding of the concepts in the chapter.

The first two chapters are written with the assumption that the reader has little or no computer experience. We wrote this workbook for a wide variety of backgrounds and disciplines so that everyone can enjoy *TheSky*. The first two chapters include detailed installation instructions for the necessary software. Computer-savvy individuals will probably want to skip these chapters and dive directly into the material.

I, Dr. Jordan, have been using *TheSky* software in my observational astronomy class for the past seven years. My students enjoy the hands-on experiences this software provides and have often expressed how essential it has been in preparing them for the activities they later do in the observatory.

When the publisher first contacted us about the Student Edition of *TheSky*, we thought there would be many sacrifices made in offering *TheSky* at a less sophisticated level. After using it, we feel that it is by far one of the best planetarium-type software products on the market today for students and amateur astronomers.

This workbook provides both instructor and student with a thorough understanding of how astronomers designate astronomical objects. Great care has been taken to explain confusing nomenclature that has been used historically and that is currently used in observational astronomy today. The exercises and questions are designed to assist students in gaining useful knowledge of how astronomical objects are named, cataloged, and found. *TheSky* provides many images that are archived for your viewing pleasure.

TheSky software package is very user-friendly. It is an excellent and fun way to learn about stars, constellations, and coordinate systems used in observational astronomy. We have discussed coordinate systems in great detail, beginning with a review of geographic coordinates. Students learn how the sky appears and how it changes for any location on Earth. Many major cities in the United States are included in *TheSky's* database, as well as many international cities and major observatories around the world. Students will enjoy taking imaginary trips to Earth's North Pole or to its Equator and all points in between.

We have thoroughly discussed both sky coordinates in this workbook. The horizon and equatorial coordinate systems are the coordinates that both amateur and professional astronomers alike use to locate objects in the sky. There are many examples and exercises are included in this workbook. They are specifically designed to use in conjunction with *TheSky*. They will help students in learning the location of objects in the sky and, most importantly, how to find them. Students will have an opportunity to change their location on the Earth and observe in real time how these changes affect the appearance and location of objects in the sky.

A troublesome concept that many students struggle with in astronomy involves the motions they observe in the sky. Many students have difficulty distinguishing true motions of Earth from apparent motions in the sky. Readers know Earth rotates on its axis and revolves around the Sun, but many may fail to understand what consequences arise from these two basic motions of Earth. This workbook discusses in depth both short-term and long-term motions of Earth and their effects on the objects in the sky.

One advantage of *TheSky* is its demonstration of the effects of earth's rotation on objects in the sky. Students and instructors can observe the continuous east-west motion of the sky over a 24-hour period. At the same instant, they can face any direction on the surface of Earth and observe this motion. Time increments (in seconds, minutes, hours, and days) can be manually set to observe short-term and long-term effects on the positions of objects in the sky. The remarkable thing is that you can make observations from *any* location on Earth.

Another interesting thing that you can do with *TheSky* is to look at the Sun's motion through the sky. This, of course, is caused by the revolution of Earth around the Sun. Users observe at first hand the location of the Sun relative to the stars and can correlate its position in the sky to the seasons on Earth. You can also observe short-term and long-term motions of the Moon and the planets. These motions puzzled many ancient astronomers for centuries and eventually led them to construct models of their "universe" that were incorrect.

TheSky provides several "sky" windows to illustrate different phenomena in the sky. The motions of the Sun, Moon, and the planets are just a few examples of interesting windows that you can open and view. The user simply opens a sky window, sits back, and observes! The authors have painstakingly prepared several multimedia windows in addition to those provided in *TheSky*. These windows are intended to help demonstrate the concepts we discuss throughout the workbook. The windows are well integrated into the workbook for continuity. Using the multimedia windows helps you gain a better insight of the phenomena.

This workbook details planetary motion and configurations. It uses *TheSky* to show their orientation within the Solar System and their appearance in the sky as well as their observed motions. This is very helpful when you are observing the planets for the first time.

The workbook also provides clear and concise explanations of planetary orientations in the sky with respect to Earth and the Sun. *TheSky* provides an excellent tool that displays three-dimensional views of the Solar System. It shows precise arrangements and alignments of all the planets for any specified time. This tool offers the user a clear understanding of where planets are located relative to Earth and Sun as well of as their appearance in Earth's sky.

Students are provided with many hands-on exercises while using *TheSky* software. The software allows users to move ahead or backwards in time. They may observe the sky as it was when Julius Caesar was in power, or, perhaps, as it will appear thousands of years in the future. By moving ahead or back in time, you can appreciate the effect of Earth's precession on the orientation of the sky and the effect this precession has on the coordinates of objects in it. This is something you would normally have to go to a planetarium to view.

Perhaps the most difficult concept to convey to students, and people in general, is that of time. A whole chapter here is devoted to time and seasons. It outlines how scientists determine time for the world, for different locations on Earth, and most importantly locally. It is certainly true that time *is relative* but it is more important to understand how your location on Earth affects the observed time of astronomical events in the sky. We have specifically designed exercises to show how one's location on Earth affects the time of observed events. Using *TheSky* allows anyone to determine the time for the beginning of the seasons for any year (past, present, or future). It helps users determine the local time of astronomical events, such as the apparent sunrise and sunset, for any location on Earth.

The most important time used by astronomers and in observational astronomy is **sidereal time**, time by the stars. Indeed, it is the time system students find most exasperating to learn and comprehend. We have taken great pains here to explain simply how astronomers keep track of this time.

Sidereal time is perhaps the most important time system in observational astronomy to master. Detailed observations are usually planned accordingly on how sidereal time is computed. The exercises provide students a comprehensive explanation of this complicated time system and how it is calculated. Time calculation is one of the fundamentals in observational astronomy that students must master in order to plan constructive observing sessions.

Several other tools are available in *TheSky* to demonstrate different types of phenomena. One interesting tool is the lunar phase calendar. It provides an up-to-date Moon calendar, for planning observation sessions.

The phasing of the Moon is a phenomenon we discuss at length in this workbook. After completing the exercises, users will have a clearer understanding of why the Moon appears the way it does in the sky. Diagrams in textbooks and on many Web pages are often confusing. Their depiction of lunar phases is difficult for many to comprehend at a glance. We have tried hard in this workbook to make sure the diagrams are correct and easy to follow. They depict the Moon's motion as well as its position relative to Earth and the Sun. In fact, after using *TheSky* and doing the exercises, you will be able to determine the moon's orbital position from simply observing its appearance in the sky. With experience, you will even be able to estimate the Moon's apparent rising and setting times as well. *TheSky* depicts lunar phases, apparent rising and setting times, and motions very accurately.

Another tool in *TheSky* is the eclipse tool. You can use it to determine the time and location of eclipses. Eclipses of the Sun and Moon are somewhat rare phenomena that people usually read or hear about after they have happened. They often occur in remote or inaccessible locations on Earth. *TheSky* allows you the luxury of looking for and observing eclipses from these mysterious and distant locations from the comfort of your own desktop.

This workbook provides a detailed description of the types of eclipses observed. It discusses the geometry of eclipses and the conditions necessary for eclipses to occur. We have clearly explained the connection between lunar phases and eclipses. In fact, with the understanding of how eclipses occur that *TheSky* provides, it helps users prepare for solar and lunar eclipses that will occur over the next few millennia.

The main objective of this workbook is to use this newly found knowledge to be able to go outside and have a greater appreciation for the wonders that many have taken for granted over the years. It is our intention to promote an easy hands-on approach to viewing the sky with *TheSky*. After completing this workbook and its exercises, students and instructors alike will become very proficient in the use of *TheSky*. The possibilities are limitless!

I (Dr. Jordan) have been associated with the teaching profession for over 25 years. I've always considered myself a "disciple of astronomy," spreading the news about what's going on in the sky above. I've been conveying these ideas and concepts to students in my astronomy classes for years. And I especially would like everyone to be able to appreciate (as I have for many years) the beauties and wonders of the sky and the Universe.

We decided to write this book to provide you, the reader, an opportunity to appreciate the wonders in the night skies as we do. With everything this software package provides, you will be astounded by the large variety of *"astronomical creatures"* that exist and the phenomena that occur in nature.

With *TheSky*, enjoy the sky as you never have before!

Tom Jordan
Scott Peters

Table of Contents

Chapter 1

Introduction

We humans have observed the stars for thousands of years. Yet today some of their names and meanings are difficult to track down. We attribute the names given to the bright stars to our ancestors to whom they had a special significance. Our forbears have also given us the names of larger groups of stars called constellations. They named constellations after mythological beasts, gods, and humans with supernatural attributes, and even common everyday objects.

The stories told long ago about these constellations rival any story told around a blazing campfire today. The ancients used these constellations to determine seasons and to remember different areas of the sky. It seems that humans can remember patterns much better than thousands of individual names. In these patterns many names have become very familiar after many years of observing.

Astronomers today continue to use the names of the constellations first plotted by the Greek astronomer Hipparchus over 2000 years ago. In 1927, by international agreement, definite boundaries were assigned to the constellations in the sky. Today astronomers still use the constellations to locate many stellar and non-stellar objects in the sky.

There are several ways to locate celestial objects. Using a star chart is perhaps the easiest. But to do so you must have some knowledge of coordinate systems used in the sky. That is what this workbook is all about. It introduces you to astronomical coordinate systems used in observational astronomy and helps you locate any naked eye or telescopic object with a click of a mouse button.

This workbook guides you through many exercises so that you can fully use the software package. *TheSky* displays the bright "naked-eye" stars and those down to the limit of the Hubble Space Telescope. It locates planets and many other Solar System objects as well as thousands of deep sky objects.

Before starting your adventure, however, you will need to successfully install *TheSky* on your PC. Listed below are the minimum system requirements necessary for the program to run effectively. Following the minimum requirements is a step-by-step installation procedure for you to follow to install the necessary files correctly.

Minimum System Requirements

TheSky program requires the following minimum system requirements:

➢ Windows 95, Windows 98, or NT 4.0 (Intel) operating system

➢ Pentium-class PC

➢ 16 megabytes of RAM

➢ 50 megabytes (MB) of free disk space (15 MB minimum)

➢ VGA monitor (Super VGA recommended)

➢ Quad-speed or faster CD-ROM drive

➢ This program *cannot* be used with Windows 3.1X or any other 16-bit operating system.

Installation of *TheSky*

It is wise to read through these installation instructions before you attempt to install *TheSky* and other software.

Windows 95/98 Instructions

1. Insert *TheSky* CD-ROM into drive.

2. The program should start automatically. If so, proceed to Step 4.

3. If the program did not start the installation automatically, then go to **Start** at the bottom left of your computer screen, then click **Run** and type **D:\setup.exe** in the dialog box. The Run window is shown in Figure 1-1. "**D**" is the designated drive letter associated with your CD-ROM drive. If "D" is not the letter of the CD-ROM drive, then substitute the correct letter for your PC, click **OK**, and go to Step 5.

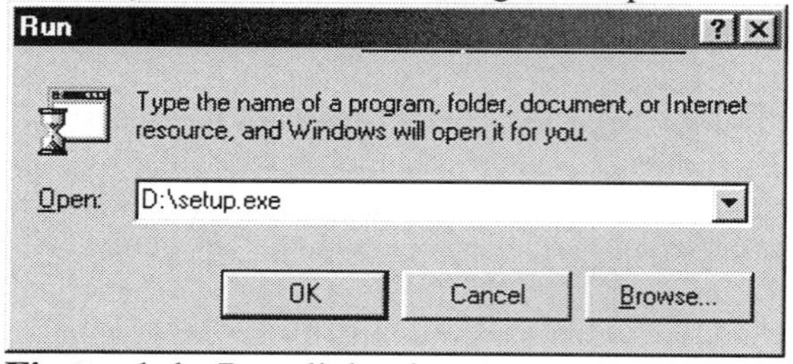

Figure 1-1 Run dialog box

4. The *TheSky* Astronomy Software Setup window will appear with the options of "Install *TheSky*" or "Cancel." This window appears in Figure 1-2. Click the Install *TheSky* button.

Figure 1-2 *TheSky* Astronomy Software Setup window

5. Several windows are displayed after you click "Install *TheSky*." The first one that appears is the Welcome window. This window provides important information before installation. The Welcome window is displayed in Figure 1-3. Read the information in the window and then click the Next button to continue.

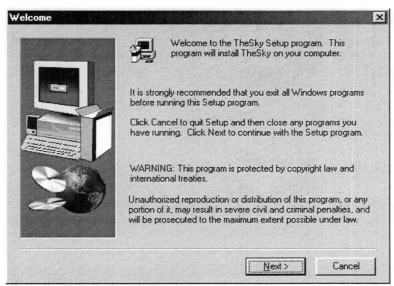

Figure 1-3 Welcome window

6. The next window that appears is the Software License Agreement window and is displayed in Figure 1-4. Please read the license agreement carefully and click the appropriate button. If you agree, then select Yes and continue with the setup. If you do not agree, then select No and you may click "Exit Setup" at this point.

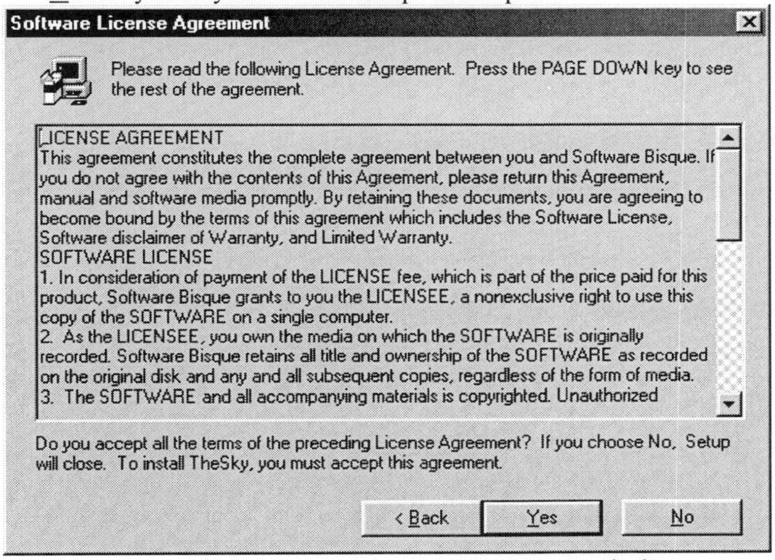

Figure 1-4 *TheSky* Software License Agreement window

7. The next window that appears is the Choose Destination Location window. This window directs you to the folder that you would like to install *TheSky*. It is displayed in Figure 1-5. If a particular folder is not chosen to install *TheSky*, the default installation will place the program into the "**C:\Program Files\SoftwareBisque\TheSky**" folder. Click the **Next** button to continue the installation.

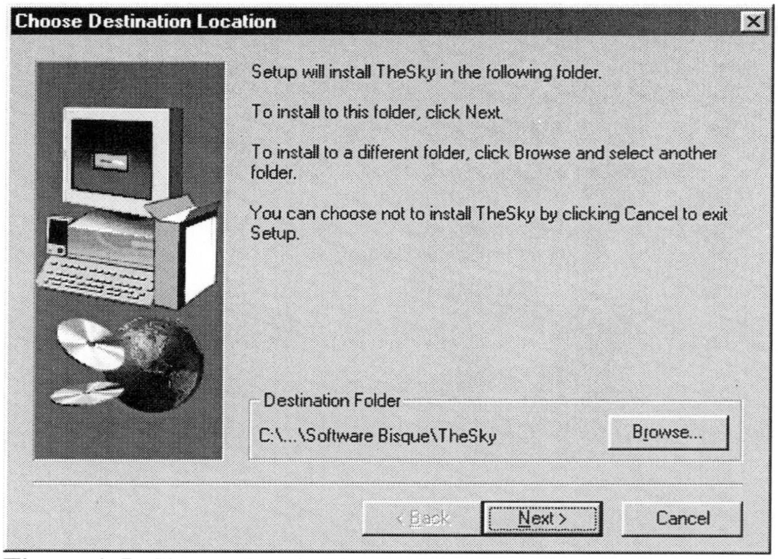

Figure 1-5 Choose Destination Location window

8. The next window that appears is the Setup Type window. This window permits you to choose the type of installation. Compact and typical installation requires the CD-ROM to be inserted into your drive before starting *TheSky*. When using the custom setup, the CD-ROM may or may not need to be present to run the program. This depends on what files are chosen to be placed on your PC's hard drive. Figure 1-6 displays the different types of installations you may choose. For this workbook, the "**Typical**" installation is highly recommended. If your disk space is limited, this installation requires only a small amount of disk drive space.

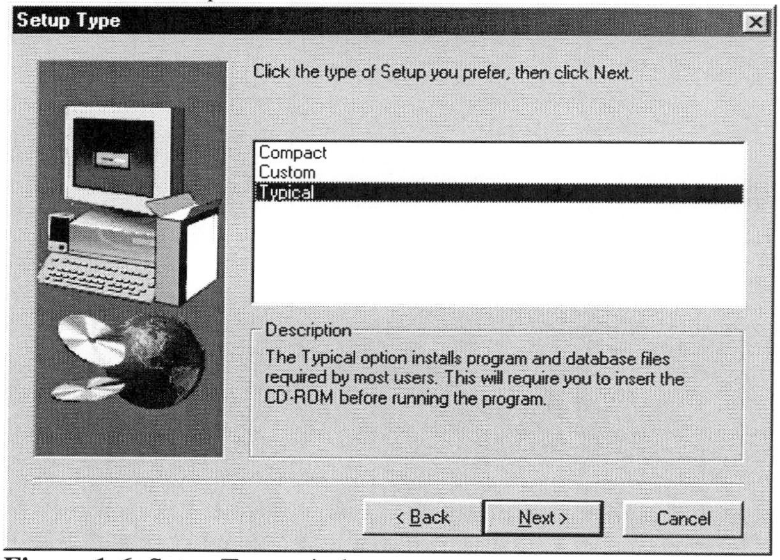

Figure 1-6 Setup Type window

9. After choosing "**Typical**," then click the **Next** button. The Start Copying Files window opens. This window displays the folder and program files being copied. The default folder is "**C:\Program Files\Software Bisque\TheSky**." This window is shown in Figure 1-7. Simply click the **Next** button to continue the installation.

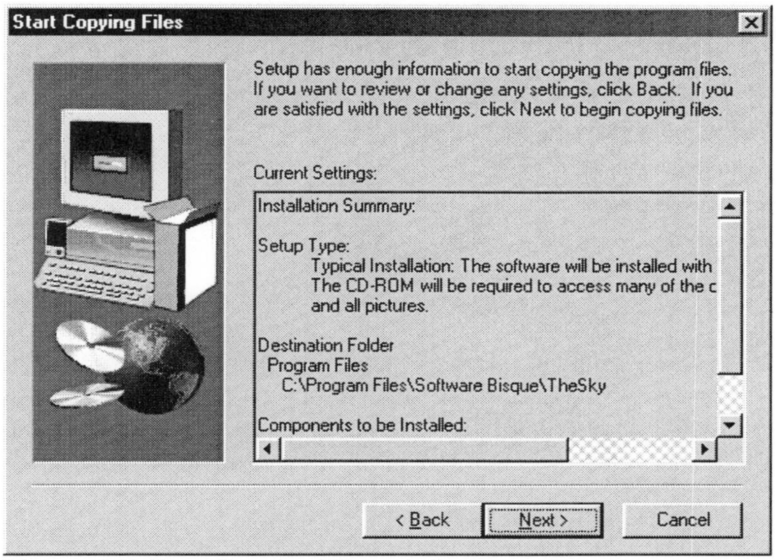

Figure 1-7 Start Copying Files window

10. After the setup program copies the necessary files to your computer, a Question window appears with the question "Would you would like to install QuickTime?" Installation of this program is purely optional! You may want to check your Program Files folder to see if QuickTime is already installed on your PC.

QuickTime 3.0.1 is a program that lets you view movies on your PC. Unless you already have QuickTime installed, you might **not** want to consider installing it. This is purely your choice!

Figure 1-8 displays the window that asks if you want to install QuickTime. If you want to install QuickTime, click **Yes**. If not, click the **No** button to continue.

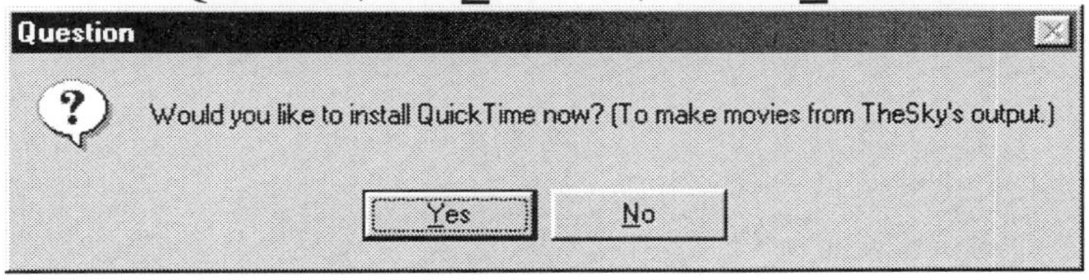

Figure 1-8 Question window

11. If you click the **Yes** button to install QuickTime, the QuickTime Setup window appears. It appears in Figure 1-9.

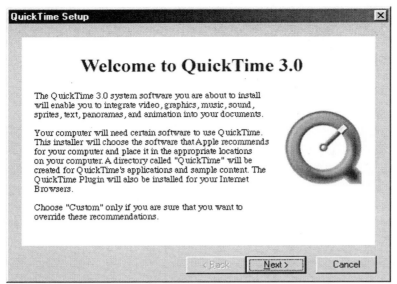

Figure 1-9 QuickTime Setup window

12. Click the **Next** button in the QuickTime Setup window.

13. After clicking the **Next** button in the window, the QuickTime 3.0 window appears, like the one in Figure 1-10.

Figure 1-10 QuickTime 3.0 window

14. Simply click the **Next** button to continue the installation of QuickTime 3.0.1.

15. The next window that appears is the QuickTime Software License Agreement. This window is displayed in Figure 1-11.

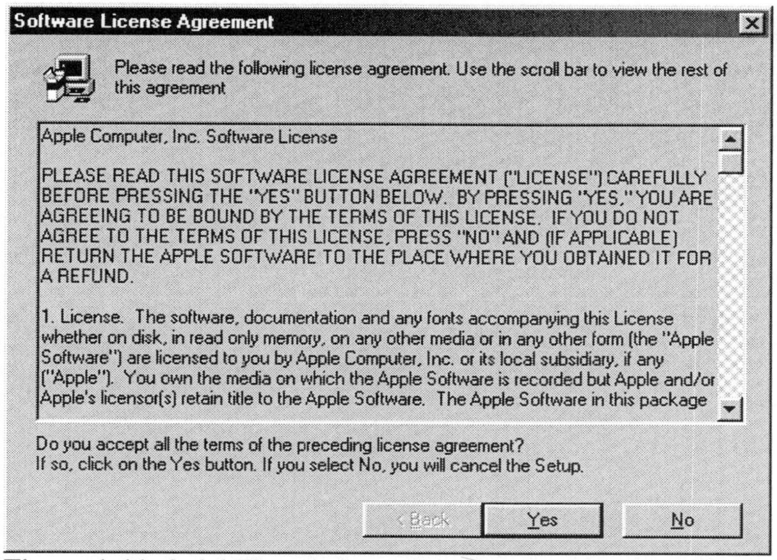

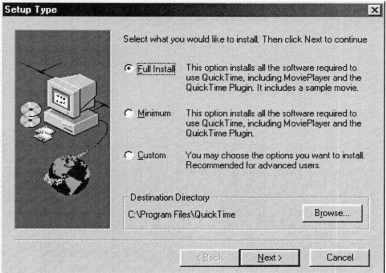

Figure 1-11 QuickTime Software License Agreement window

16. If you agree to the software license statement, then click the **Yes** button to continue. If not, click the **No** button and exit the setup.

17. After you click the **Yes** button the Setup Type window then appears. It is displayed in Figure 1-12.

Figure 1-12 Setup Type window

18. We recommend that you choose "Full Install" (full installation) of the QuickTime software and click the **Next** button.

19. The next window that opens in the installation is the QuickTime Plugin Options window. It is advisable to install both the Netscape Navigator and the Microsoft Internet Explorer plugins if you have both these programs installed on your PC. You must have at least one of these programs installed before you can continue with the installation. The window for the plugin options is displayed in Figure 1-13.

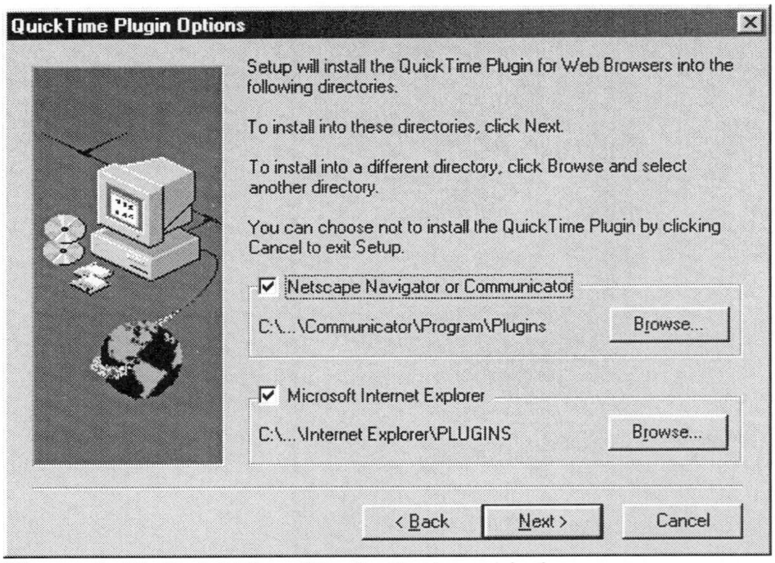

Figure 1-13 QuickTime Plugin Options window

20. Once you have chosen the plugin(s) you want installed, click the **Next** button. This displays the Create or Select Program Folder window for the QuickTime installation. The folder highlighted is labeled QuickTime, as displayed in Figure 1-14.

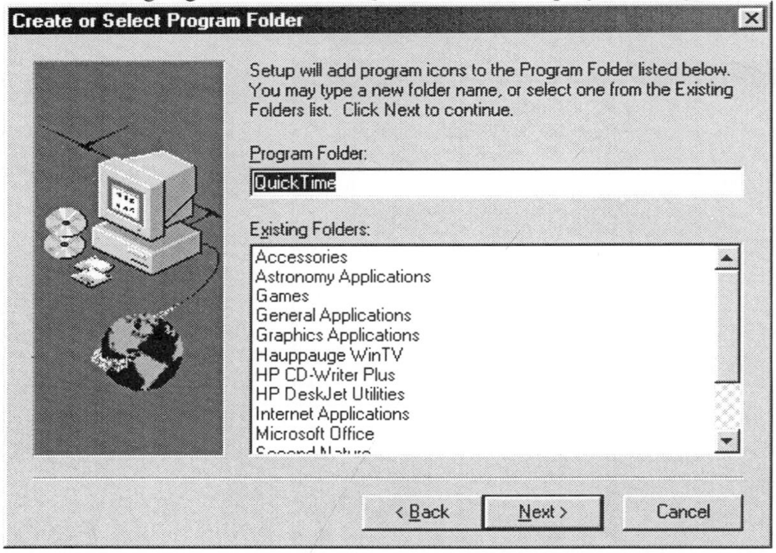

Figure 1-14 QuickTime Create or Select Program Folder window

21. At this point, when the QuickTime Create or Select Program Folder window appears you only have to click the **Next** button to continue.

22. The last window that appears during the installation is the QuickTime Setup Finish window, which is shown in Figure 1-15.

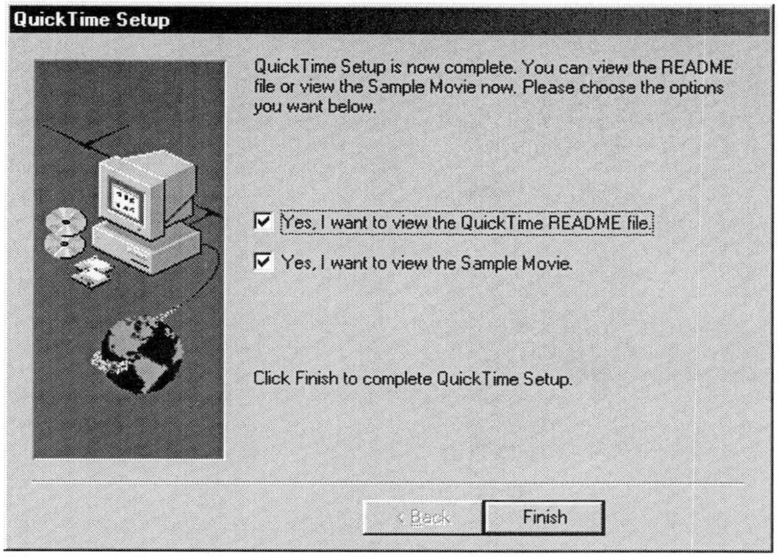

Figure 1-15 QuickTime Setup finish window

23. Click the **Finish** button and read the README file if you wish to ascertain important information about technical support, upgrades and the operating manual for QuickTime.

    Note: If you decide not to install QuickTime on your PC, you may do so later by inserting the CD-ROM into your PC and follow Steps 10 through 23 once again.

    To save **your** city's latitude and longitude or **any** other location that is _not_ in *TheSky* database, you will need to change one of the file settings in *TheSky*. The following procedure outlines how to change this setting by using Windows Explorer.

Saving Your City's Latitude and Longitude

1. Click on the Start button. ()

2. Go to **Programs**, and then to **Windows Explorer**, as shown in Figure 1-16.

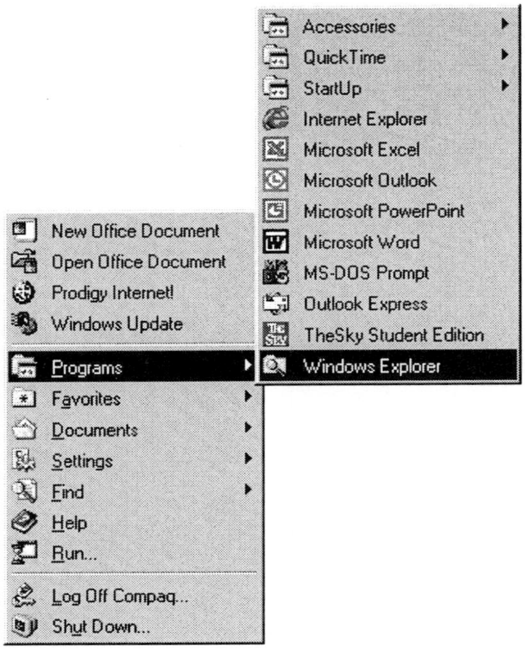

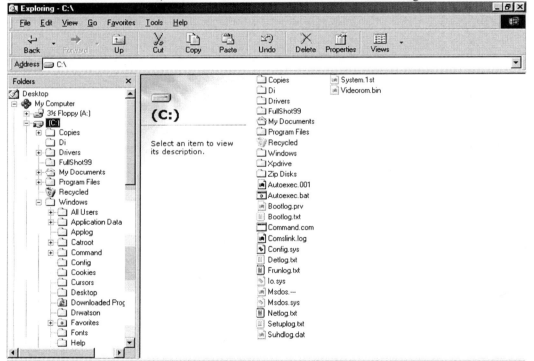

Figure 1-16 Programs menu

3. In the left-hand window, double-click on the C-drive as illustrated in Figure 1-17.

Figure 1-17 C-drive as displayed in Windows Explorer

4. Double-click on the **Program Files** folder as displayed in Figure 1-18.

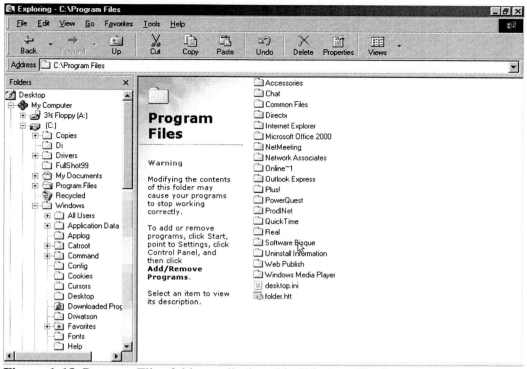

Figure 1-18 Program Files folder as displayed in Windows Explorer

5. Continue by double clicking on the **Software Bisque** folder as illustrated in Figure 1-19.

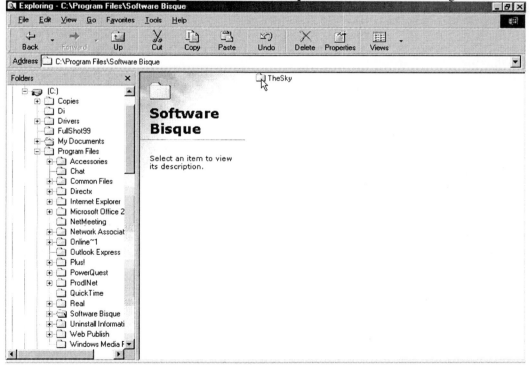

Figure 1-19 Software Bisque Folder as displayed in Windows Explorer

6. Double-click on the *TheSky* folder as shown in Figure 1-20.

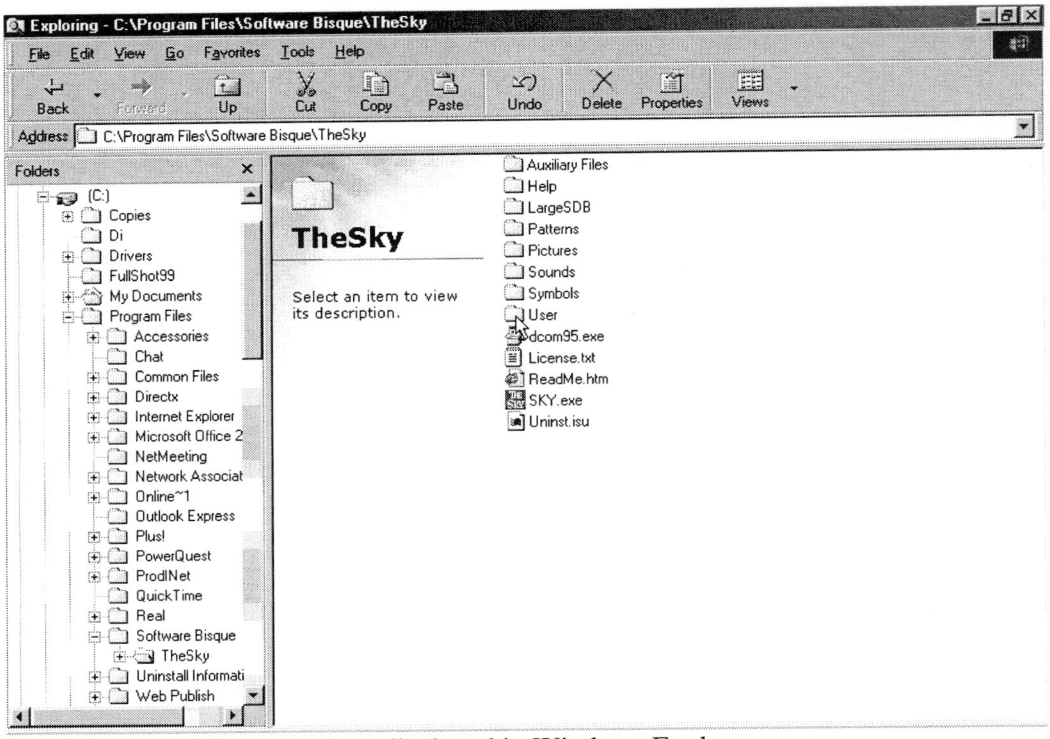

Figure 1-20 – *TheSky* folder as displayed in Windows Explorer

7. Double-click on the **User** folder as seen in Figure 1-21.

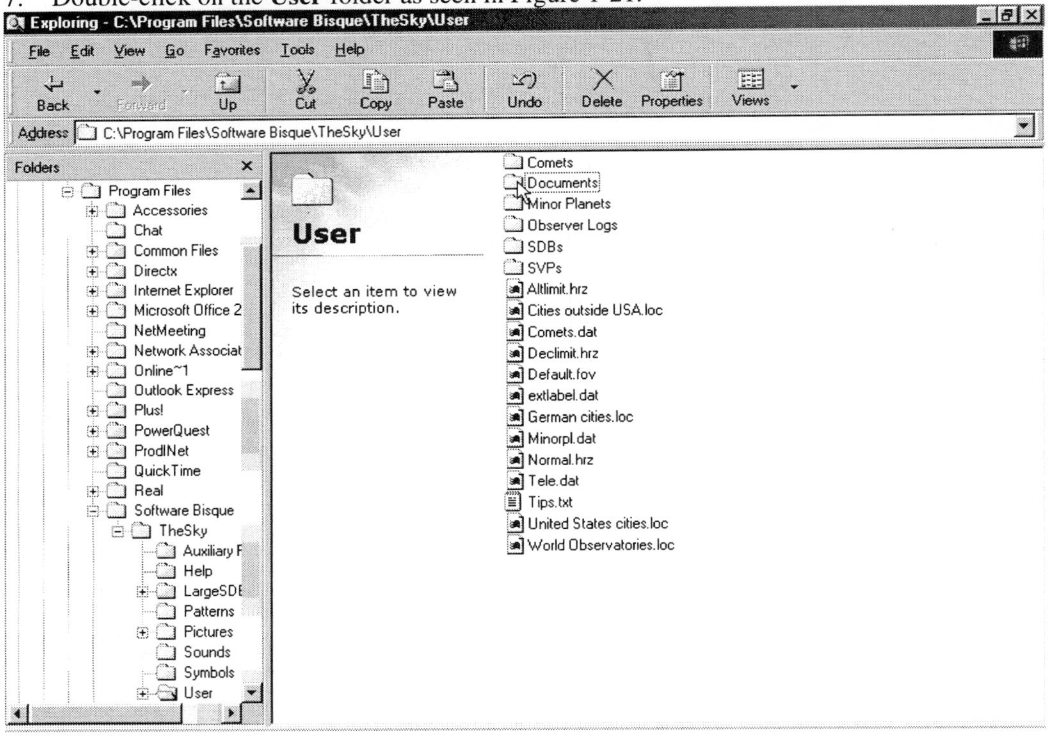

Figure 1-21 User folder as displayed in Windows Explorer

8. Click on the file named **United States cities.loc** using the right mouse button (right-click). This opens the menu shown in Figure 1-22.

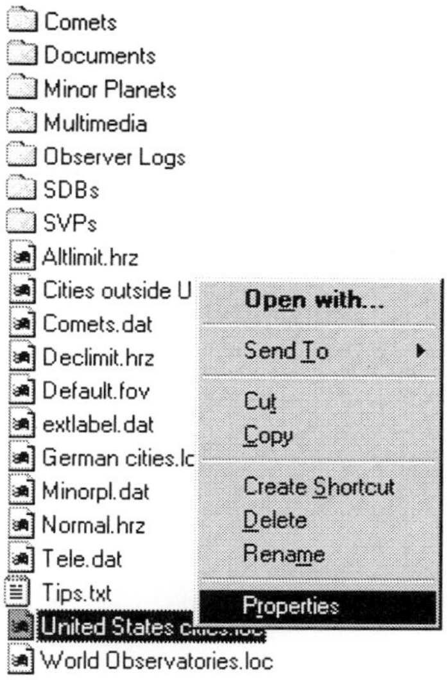

Figure 1-22 Windows Explorer menu

9. Click on "**Properties**." This window is displayed in Figure 1-23.

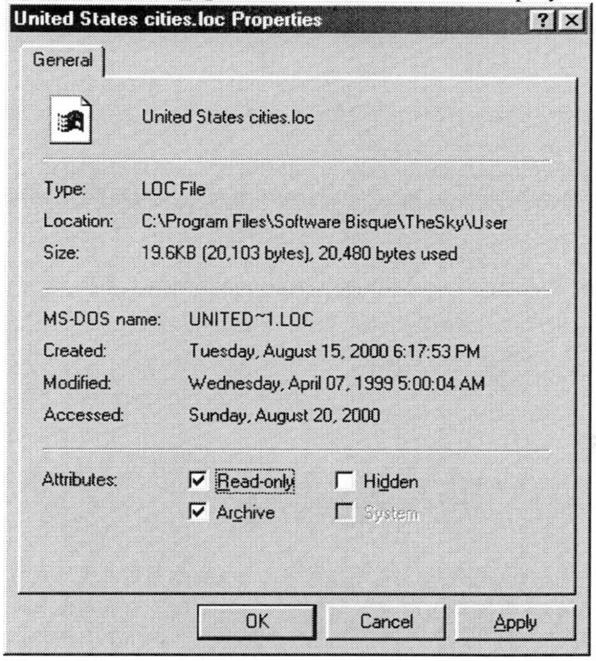

Figure 1-23 United States cities.loc Properties window

10. Click the **Read-only** box in the Attributes section of this window. To do this, simply click the mouse on the checkmark. This will turn off this option, which is necessary to save locations other than those in *TheSky's* database. The window looks like the one shown in Figure 1-24.

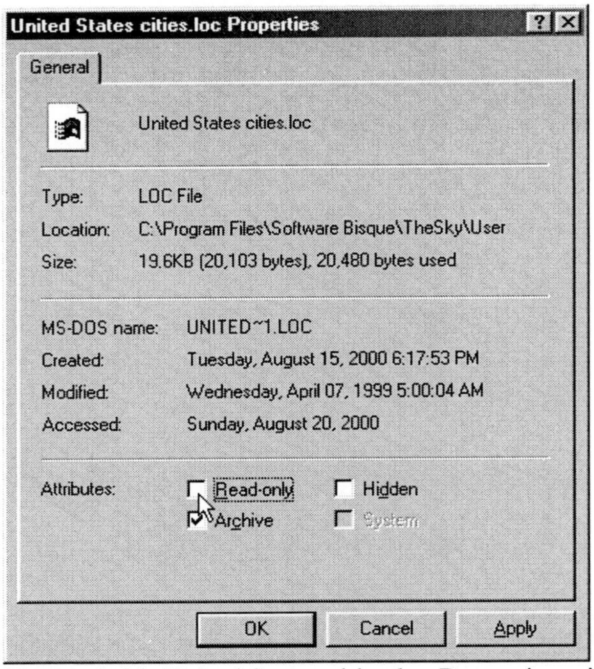

Figure 1-24 United States cities.loc Properties window

11. Click OK.

12. Do **not** close Windows Explorer yet. There is one more task that needs to be performed before using *TheSky*.

To use the multimedia files discussed in later chapters, you will need to create a file folder entitled Multimedia within *TheSky* folder, by using Windows Explorer. The following procedure describes how to create this folder.

Creating a Multimedia File Folder

1. Double-click on the **Documents** folder again as shown in Figure 1-25.

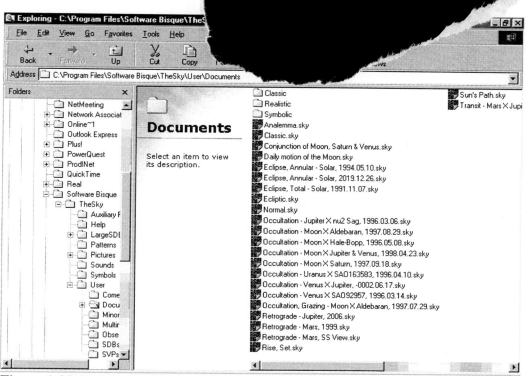

Figure 1-25 Documents folder as displayed in Windows Explorer

2. Click the **File** button on the menu bar, and go to **New**, then **Folder**. This is displayed in Figure 1-26.

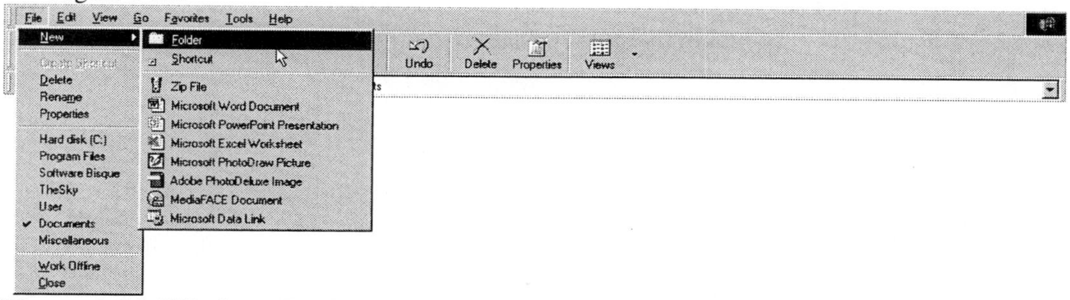

Figure 1-26 Windows Explorer menu

3. Type the word **Multimedia**, then hit the "**Enter**" key.

4. Your Multimedia folder has now been created. The complete path for the Multimedia folder should be

C:\Program Files\Software Bisque\User\Documents\Multimedia

5. Later on you need to access and download multimedia files that you will use with this workbook from the Web page of Brooks/Cole Publishing Company. The URL is: http://www.brookscole.com/product/0534390722.

6. You are now ready to start using *TheSky*.

The purpose of this chapter is to help users become familiar with the various toolbars in *TheSky* software. Before you start any exercise in this workbook or use *TheSky* on your own, we need to explain what the toolbars at the top of the sky window mean and, more importantly, what they do! They are there to assist you in making shortcuts in the program. Several toolbars are displayed in *TheSky*.

The *Standard toolbar* provides several options within the current opened sky window. With it, the user can open archived windows or save current window settings of the sky. You can even print out the current sky window with this toolbar.

The *Orientation toolbar* helps in orienting the sky and the direction in which the user is looking. It allows you to zoom in or out on a particular region in the sky.

The *View toolbar* provides shortcuts to finding the coordinates of celestial objects. It allows users to display times that objects appear to rise, set, or transit the local celestial meridian. It lets them change their view of the sky from the night sky mode to a day sky mode and vice versa. It allows the toggling on and off of names of objects and constellation lines and boundaries. It will even let users change the sky window to a **night vision** mode.

The *Objects toolbar* lets users view a variety of nonstellar objects. The types of objects can be selected and displayed in the sky window. It also allows users to find objects in the sky.

The *Time Skip toolbar* permits incrementing time steps while viewing the sky. With this toolbar, you can view changes in the positions of objects over short or long periods of time. We discuss detailed information about these toolbars in the following pages.

Using the Standard Toolbar

The first toolbar is the *Standard toolbar*. It contains buttons for creating new documents, opening existing documents, saving active documents, copying, pasting, printing information about *TheSky*, and getting help. The icons in this toolbar and their function in the software are displayed in detail in the following list:

  Button creates a new document in the current window.

  Opens an existing document in the current window.

  Saves the document displayed in the current window.

  Copies the star chart to the clipboard.

Pastes the object that is being stored on the clipboard.

Prints the active window as a star chart (white background and black objects).

 Displays information about *TheSky* and your computer.

Displays help information on clicked buttons, menus, and windows.

Using the Orientation Toolbar

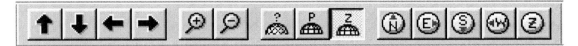

The *Orientation toolbar* contains many buttons. These buttons control where the user is looking in the sky. With them, you can move the sky left, right, up, and down. In addition to these directions, it permits the user to select different orientations of the sky. The yellow buttons control the direction in which *TheSky* is plotted on the computer monitor and the direction that the user is facing.

 Scrolls the display down, but moves up from the ground.

 Scrolls the display up, but moves down toward the ground.

 Scrolls the display right, but moves the ground to the left.

 Scrolls the display left, but moves the ground to the right.

 Zooms in on an object, and decreases the field of view.

 Zooms out on an object, and increases the field of view.

 Allows user-defined rotation of the sky.

 Orients the sky with the celestial pole upward.

 Orients the sky with the zenith upward.

 Changes display so user is looking north.

-17-

Changes display so user is looking east.

 Changes display so user is looking south.

 Changes display so user is looking west.

 Changes display so user is looking up at the zenith.

Using the View Toolbar

The *View toolbar* contains buttons to configure *TheSky* in a variety of ways. It allows users to view the Solar System in a three-dimensional mode. It lets users view the sky in a daytime or nighttime mode. It also allows users to display the equatorial or horizon grid onto the sky. These grids are useful in determining where astronomical objects are located in the sky. You can display constellation lines and boundaries using this toolbar. Several other screen-viewing options are also available to users.

 Shows the Solar System in a three-dimensional mode.

 Shows daytime/nighttime sky.

 Displays or hides the equatorial grid.

 Displays or hides the horizon grid.

 Displays or hides the constellation names and lines.

 Displays or hides the constellation boundaries.

 Displays or hides the common names of objects.

 Changes the color scheme of the screen in order to preserve the user's night vision.

Displays or hides the caption, menu, and toolbars.

Using the Objects Toolbar

The *Objects toolbar* has just a few buttons on it. It lets users find many astronomical objects in *TheSky* database. Nonstellar objects such as galaxies, star clusters, and nebulae can be toggled on and off with a single mouse click.

 Allows user to find any astronomical object.

 Displays or hides the stars.

 Displays or hides galaxies.

 Displays or hides clusters.

 Displays or hides nebulae.

Using the Time Skip Toolbar

The *Time Skip toolbar* allows users to control or input increments of time that can be skipped either ahead or behind in the sky. It also controls tracking features in the software. Users can observe short-term motions of objects such as the east–west movement of the Sun, Moon, planets, and stars. Users can also observe long-term motions of the Sun, Moon, and planets.

The window displayed as follows controls the input time increment. You can set this increment to 1 second, 1 minute, 1 hour, 1 day, sunrise, sunset, a synodic day, or a sidereal day. In fact, you can use any time you want.

 Resets the time and location settings.

Begins a backward time step.
The time step occurs rapidly, so be prepared!

 Initiates one backward time step.

 Stops the time skip.

 Initiates one forward time step.

 Begins a forward time step (play).
The time step occurs rapidly, so be prepared!

 Records an object's time-skipped trails.

 Displays Solar System tracking setup box.

Now try the exercises on the following pages.

Chapter 2

TheSky Review Exercises

TheSky Exercise 1: Setting Your Location, Date, and Time in *TheSky*

1. Click "Data" at the top of the sky window, then "Site Information." Now click the Location tab. Make certain that the information displayed on your monitor is identical to that shown in Figure 2-1.

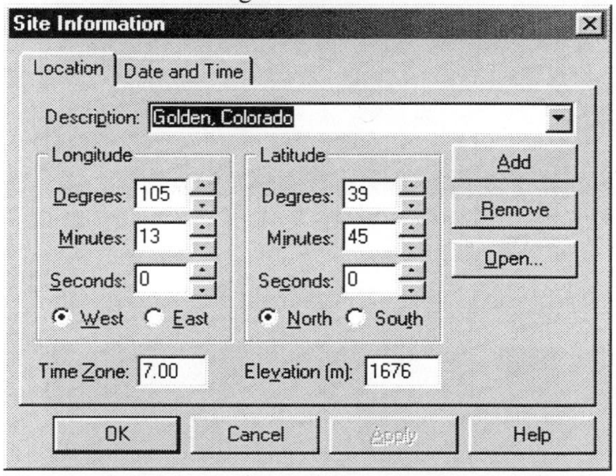

Figure 2-1 Site Information window for Golden, Colorado

2. If the information is not identical, then click the pull-down arrow on the Description line. Scroll down the menu and click on "Golden, Colorado."

3. Now click on the Date and Time tab and then on the box labeled "Use computer's clock." Now set the date to January 1, 2003 and the time to 20:00. *Note: TheSky* uses a 24-hour clock instead of a 12-hour clock so 20:00 is 8 P.M. The window now looks like the one illustrated in Figure 2-2. Click OK.

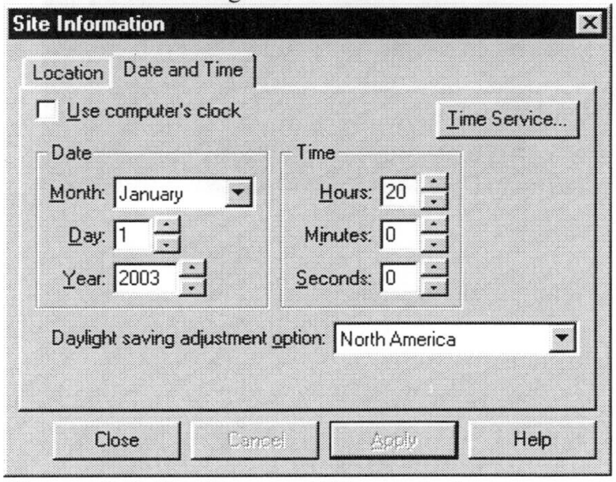

Figure 2-2 Set Date and Time window

4. Click the South shortcut button (⑤) located on the Orientation toolbar. Your screen now looks like the one shown in Figure 2-3.

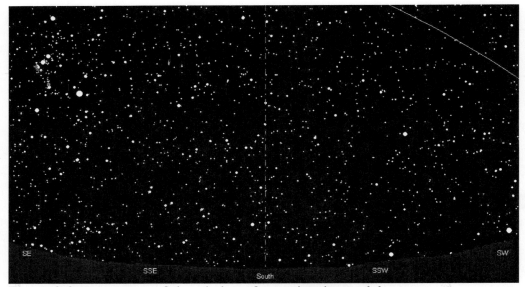

Figure 2-3 Appearance of sky window after setting time and date

5. Click the stars shortcut button (⬛) on the Objects toolbar. Your screen will appear like the one in Figure 2-4.

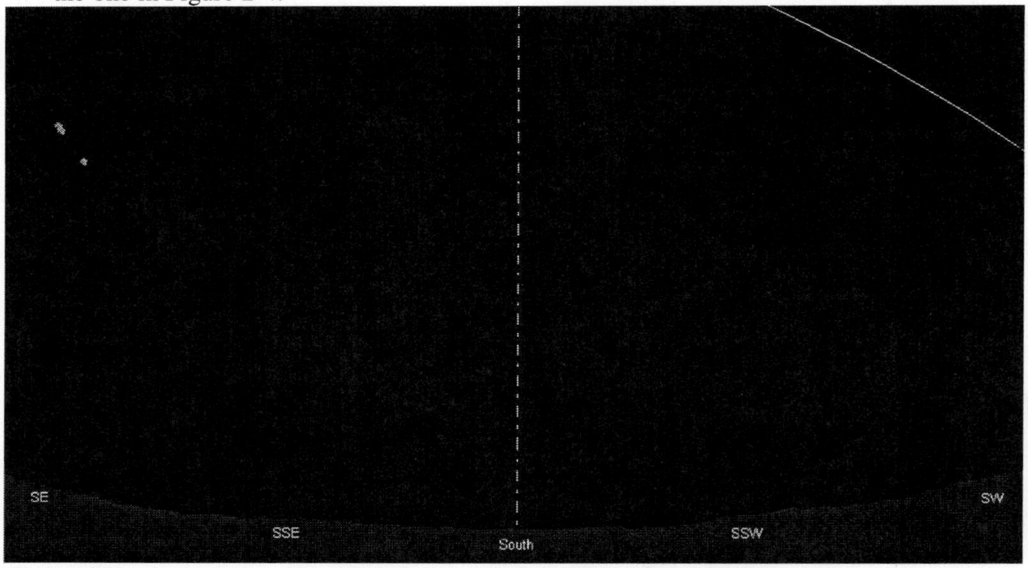

Figure 2-4 Appearance of the sky window after clicking off the stars

6. If you click the ⬛ button again, the stars return to the sky window.

7. Now try using some of the other buttons on your own.

TheSky Exercise 2: Changing Your Location, Date, and Time in *TheSky*

1. Click "Data" at the top of the sky window, then "Site Information." Click the Location tab once again.

2. Now click the pull-down arrow on the Description line. Scroll down the menu and click on "Muncie, Indiana." Click "Apply" then "Close."

3. Make certain that the information displayed on your monitor is the same as the information shown in Figure 2-5.

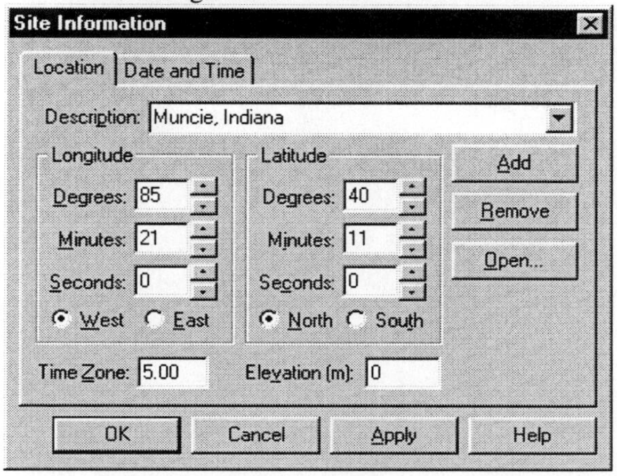

Figure 2-5 Changing one's location to Muncie, Indiana

4. Now click on the Date and Time tab to set the date to April 25, 2015, and the time to 21:00. The window now looks like the one displayed in Figure 2-6. Click "Apply," then "Close."

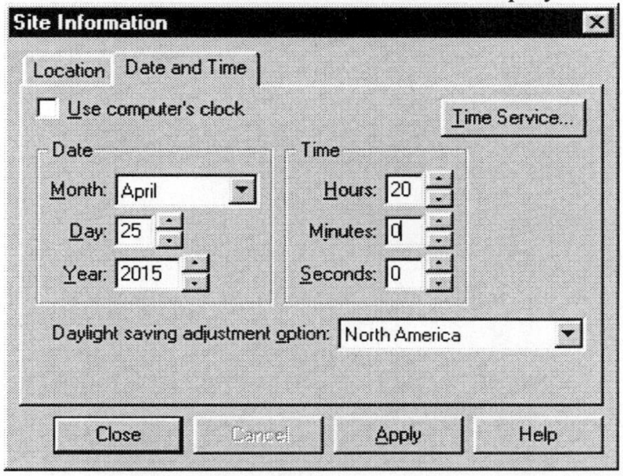

Figure 2-6 Setting the date and time to April 25, 2015, at 21:00

5. Click the South shortcut button () located on the Orientation toolbar.

6. Click the up blue button () three times. The screen now looks like the one illustrated in Figure 2-7. Notice that the constellation Leo is due south in this sky window.

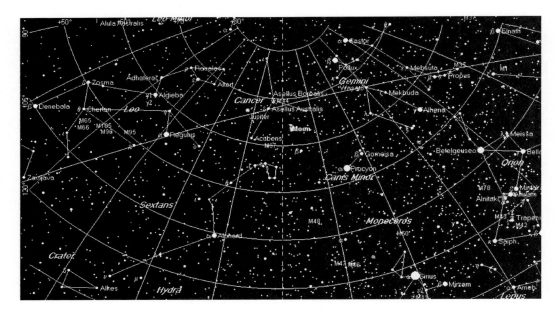

Figure 2-7 The sky looking south on April 25, 2015, from Muncie, Indiana

7. Now click the "Common Names" () button on the View toolbar. Notice that the names of the objects disappear. The sky window looks just like the one in Figure 2-8.

Figure 2-8 The sky looking south after toggling off the Common Names button

8. Notice that only the names disappeared, not the objects.

9. Click the ( ) button again and the labels return to the sky window.

Now, some of the features in the *Normal.sky* window file have been changed since you started these exercises. *A word of caution is warranted here!* On exiting the program, **if any** changes have been made in the Normal.sky window, the program prompts the user by displaying a window like the one in Figure 2-9. It asks "Save the changes to the Normal.sky?"

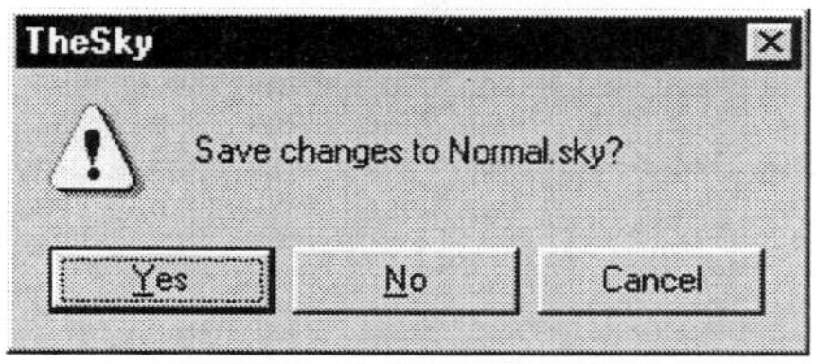

Figure 2-9 The Save Changes to Normal.sky window

Until you have become familiar with this software package, we recommend that you click **No** to the prompt in Figure 2-9. The next time *TheSky* is run, however, you must reset the location to your location. If you save the Normal.sky file at any time, it will always open to the current sky window settings, the ones you have previously made.

Once your location and personal preferences have been set for *TheSky* window, then **save** the window file. We recommend that you make a backup copy of the Normal.sky file. You may copy this file to a floppy disk or to another folder on your PC. If you wish to keep this file *TheSky* program folder it must be renamed. It is appropriate to name this backup file as **Normal.bak**. Then, it may be copied to the following folder:

C:\Program Files\Software Bisque\TheSky\User\Documents

Just be sure to keep this file in a safe place for future use.

Chapter 3

Setting Your Location and Time in *TheSky*

The purpose of this chapter is to help users set their site information and other defaults in *TheSky*. *TheSky* uses your latitude and longitude to accurately display objects that can be seen from any location on Earth. This information, of course, is different for each observer. Some major cities have been included, but it is best to use *your* latitude and longitude before starting.

Exercise A: Setting the Site Information

1. Run *TheSky*.

2. Click "Data" on the toolbar at the top of the sky window.

3. Click on "Site Information." This opens a window like the one shown in Figure 3-1.

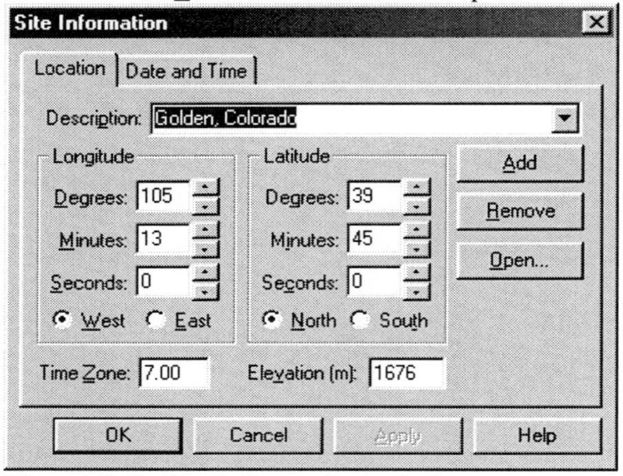

Figure 3-1 Site Information window

4. Make certain the Location tab is selected first.

5. The current city listed in the box labeled Description is Golden, Colorado. This is the default setting in *TheSky*.

6. Delete the words "Golden, Colorado."

7. Type a descriptive name (no more than 30 characters) for your location. In this exercise, type in "Ball State Observatory."

8. In the area entitled "Longitude" and "Latitude" the coordinates will be changed to those of the Ball State Observatory.

9. 85° 21′ 38″ W is the longitude for the observatory. Type the number 85 in the Degrees box, a 21 in the Minutes box, and a 38 in the Seconds box. Because the observatory is located west of Greenwich, England click "West."

10. 40° 11′ 58″ N is the latitude for the Observatory. Type the number 40 in the Degrees box, an 11 in the Minutes box, and a 58 in the Seconds box. Because the observatory is located north of the geographic equator, click "North."

11. Enter the time zone and elevation (in meters) of your current location. The time zone at the Ball State Observatory is +5.00, and the elevation is approximately 322 meters. The Site Information window looks like the one shown in Figure 3-2.

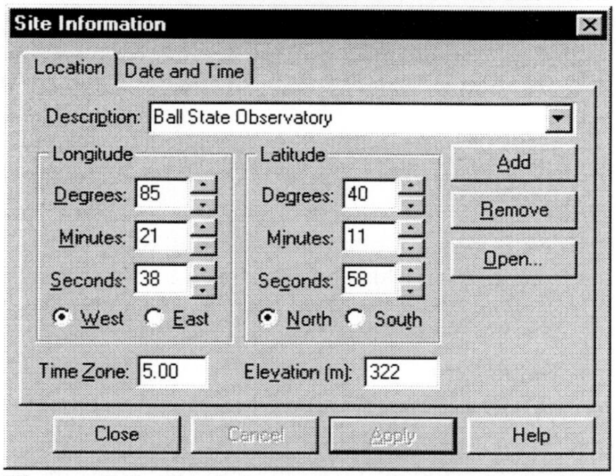

Figure 3-2 Site Information window for Ball State Observatory

12. If your location is *not* listed in the indexed menu, then you may click on the "Add" button at the right to save it to *TheSky's* database. Click "Add" to save Ball State Observatory's location in your database.

13. Click "Apply." The sky automatically changes the desktop display to your location.

14. Now click on the Date and Time tab.

15. There are two time modes in which *TheSky* operates in the Date and Time menu. The first mode uses the current time of your computer's clock. The second mode is one that allows you to set the date and time manually.

16. Click the box labeled "Use computer's clock." This will "turn off" this mode and allows the time and date to be entered manually.

17. Set the date to December 31, 2005.

18. Set the time to 9:00 P.M. On a 24^H clock, this is $21^H 00^M$. The Site Information window now looks like the one displayed in Figure 3-3.

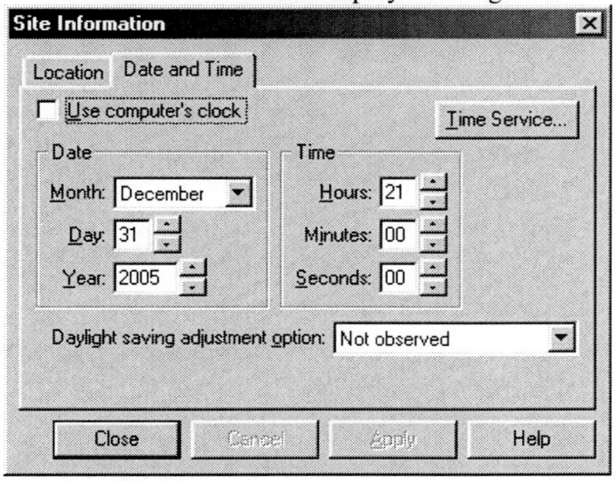

Figure 3-3 Site Information window showing date and time

19. A change in the time may be necessary if your location does not observe Daylight Savings Time. *TheSky* software assumes that you do. The default setting in the program is "North America."

   If your location does not observe it, then this option must be changed to "Not observed." Click on the menu arrow and scroll to this option. This option is located at the top of the indexed list.

20. Click "Apply," then "Close."

21. Your location has now been changed from Golden, Colorado, to the Ball State Observatory, located in Muncie, Indiana. The date and time have also been changed to December 31, 2005, at 9:00 P.M.

Exercise B: Let's find an object in *TheSky*

1. Click Edit on the toolbar at the top of the sky window, then Find, or simply type the letter "**F**." The Find window appears like the one illustrated in Figure 3-4.

Figure 3-4 Find window

2. Find Messier Object Number 42 (M42).

3. To do this, click on "Messier Objects" in the Common Names box. Figure 3-5 displays the Common Names list with the Messier Objects highlighted.

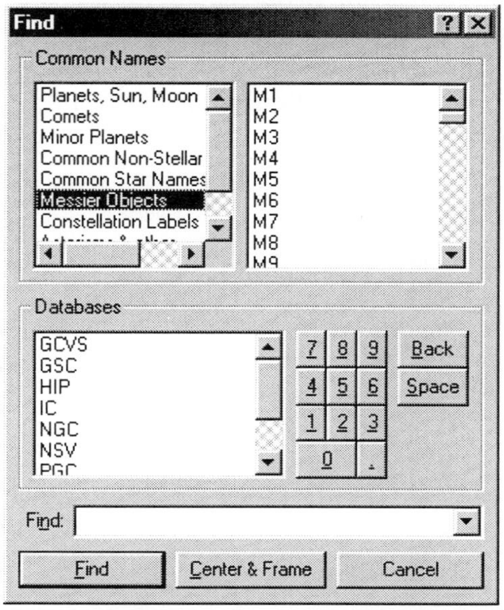

Figure 3-5 Find window displaying Messier Objects

4. Scroll down the list of objects indexed in the right-hand box until you find "M42," then click on it. Figure 3-6 displays the "Find" window for M42.

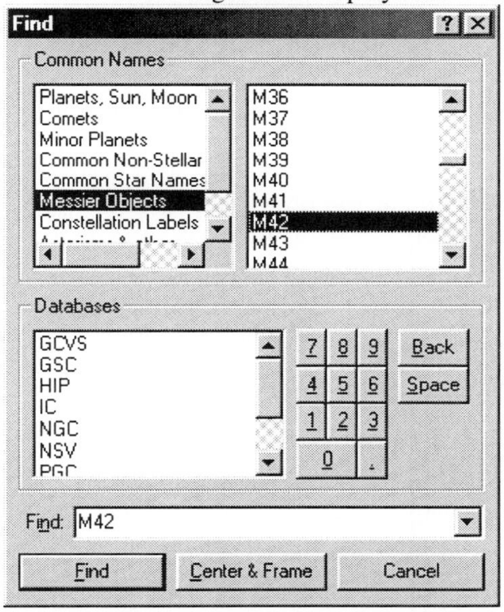

Figure 3-6 Find window displaying M42

5. Click the Find button (Find).

6. The window shown in Figure 3-7 appears.

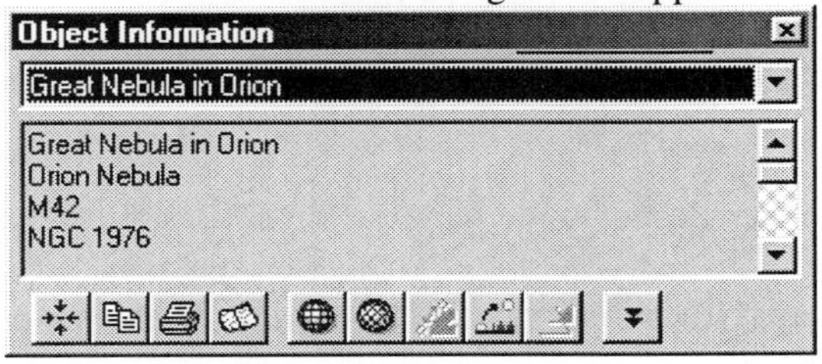

Figure 3-7 Object Information window for M42

7. Click the "More Information" button (⬇).

8. The Object Information window now appears, like the one displayed in Figure 3-8.

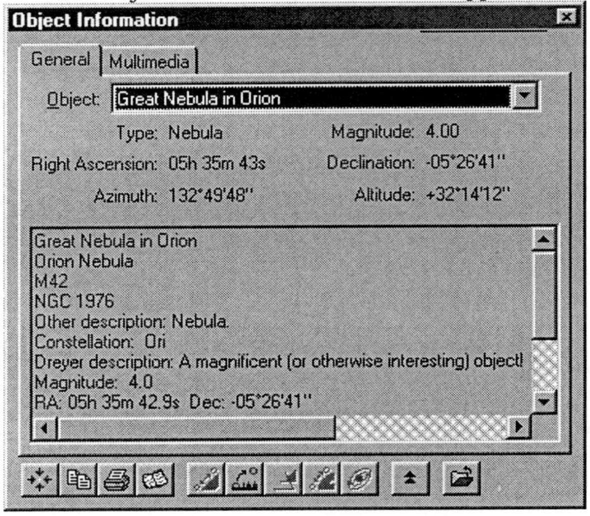

Object Information

General | Multimedia

Object: Great Nebula in Orion ▼

Type: Nebula   Magnitude: 4.00
Right Ascension: 05h 35m 43s   Declination: -05°26'41"
Azimuth: 132°49'48"   Altitude: +32°14'12"

Great Nebula in Orion
Orion Nebula
M42
NGC 1976
Other description: Nebula.
Constellation: Ori
Dreyer description: A magnificent (or otherwise interesting) object!
Magnitude: 4.0
RA: 05h 35m 42.9s Dec: -05°26'41"

Figure 3-8 Object Information window after clicking ⬇

9. Click the "Center" button (✶).

10. Write down the azimuth and altitude of M42 from the Object Information window as shown in Figure 3-9 for this location.

Azimuth = _____ °   Altitude = _____ °

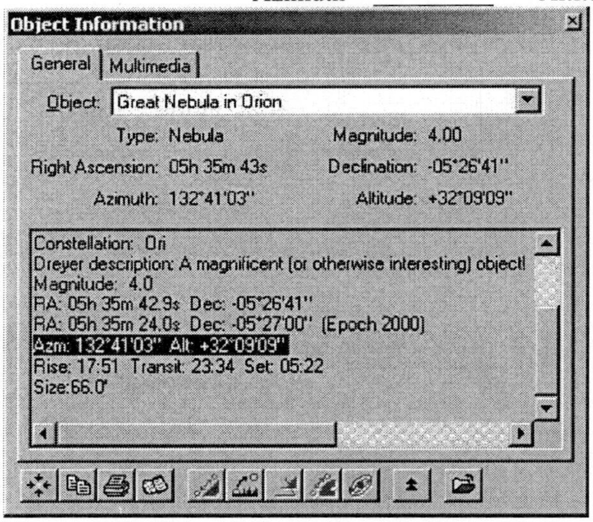

Object Information

General | Multimedia

Object: Great Nebula in Orion ▼

Type: Nebula   Magnitude: 4.00
Right Ascension: 05h 35m 43s   Declination: -05°26'41"
Azimuth: 132°41'03"   Altitude: +32°09'09"

Constellation: Ori
Dreyer description: A magnificent (or otherwise interesting) object!
Magnitude: 4.0
RA: 05h 35m 42.9s Dec: -05°26'41"
RA: 05h 35m 24.0s Dec: -05°27'00" (Epoch 2000)
Azm: 132°41'03" Alt: +32°09'09"
Rise: 17:51 Transit: 23:34 Set: 05:22
Size: 66.0'

Figure 3-9 Object Information window with the azimuth and altitude of M42

11. Close the Object Information window by clicking on the "x" button (⊠) in the upper right-hand corner of the window.

12. *TheSky* window looks exactly like the one illustrated in Figure 3-10. M42 is the circled object in the figure.

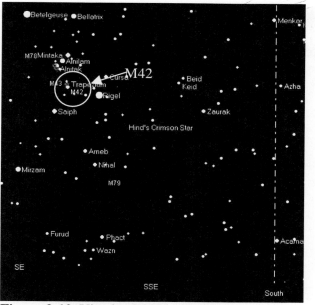

Figure 3-10 Viewing M42 in *TheSky*

There is another way to find M42 or any other object in the sky.

Exercise C: Let's find another object in *TheSky*

1. After you click on Edit on the toolbar at the top of the sky window, then Find, the Find window reappears as before. This time, instead of clicking on "Common Names," click on the Find description line located near the bottom of the Find window. The cursor will blink after you have clicked on it.

2. Type"M45" in the description line. The Find window appears, like the one illustrated in Figure 3-11.

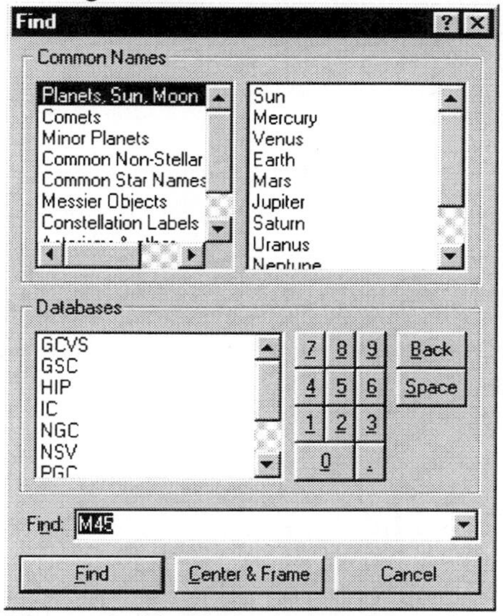

Figure 3-11 Find window for M45

3. Now click the Find button ([Find]) and the Object Information window for M45 appears. It is displayed in Figure 3-12.

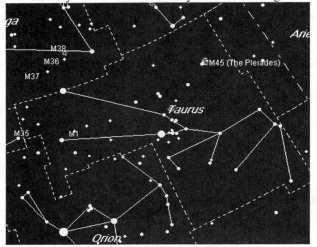

Object Information

General | Multimedia

Object: M45 (The Pleiades)

Type: Mixed Deep Sky   Magnitude: 0.00

Right Ascension: 03h 47m 54s   Declination: +24°09'37"

Azimuth: 286°44'00"   Altitude: +18°09'06"

```
M45 (The Pleiades)
30
Sky Database: Messier Objects
RA: 03h 47m 53.7s  Dec: +24°09'37"
RA: 03h 47m 00.0s  Dec: +24°07'00"  (Epoch 2000)
Azm: 286°44'00"  Alt: +18°09'06"
Rise: 06:44  Transit: 14:15  Set: 21:47
```

Figure 3-12 Object Information window for M45

4. Click the "Center" button ( ) to center the Pleiades star cluster on your computer screen. The sky window looks exactly as shown in Figure 3-13.

Figure 3-13 Star Cluster M45 in Taurus

The user can find any object in *TheSky* database by following the previous exercises. The second method saves you time instead of clicking the "Common Names" and scrolling through the indexed objects.

The fastest way to find astronomical objects in *TheSky* database is to right click your mouse anywhere in the sky window. A small menu appears, allowing users to access the Find window and locate objects quickly and easily. This menu enables the user to perform a variety of other functions in *TheSky* such as changing the preferences and filters in your sky window.

Exercise D: Let's find M42 from Miami, Florida

1. Click "Data" on the toolbar at the top of the sky window.

2. Click on "Site Information." This opens a dialog box like the one in Figure 3-14.

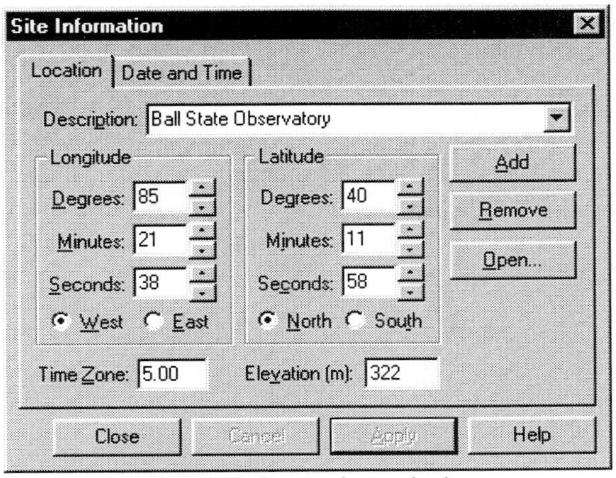

Figure 3-14 Site Information window

3. Click on the down arrow (⏷) in the "Description" box.

4. Select "Miami, Florida" from the menu of indexed cities. The Site Information window now looks like the one displayed in Figure 3-15.

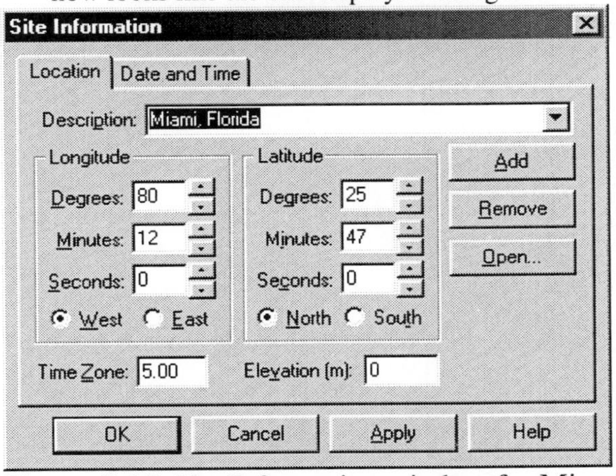

Figure 3-15 Site Information window for Miami, Florida

5. Click OK ( OK). (Do *not* change the date and time.)

6. Now find M42 (Review Exercise A or B).

7. Write down its azimuth and altitude from this location.

Azimuth = _____°   Altitude = _____°

Exercise E: Let's find M42 from Golden, Colorado

1. Click "Data" on the toolbar at the top of the sky window.

2. Click on "Site Information." This opens a window like the one shown in Figure 3-16.

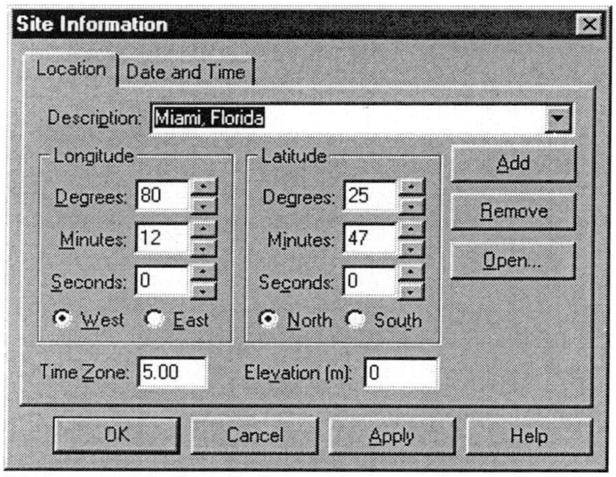

Figure 3-16 Site Information window

3. Click on the down arrow () in the "Description" box.

4. Scroll down the list and select "Golden, Colorado." The Site Information window now looks like the one displayed in Figure 3-17.

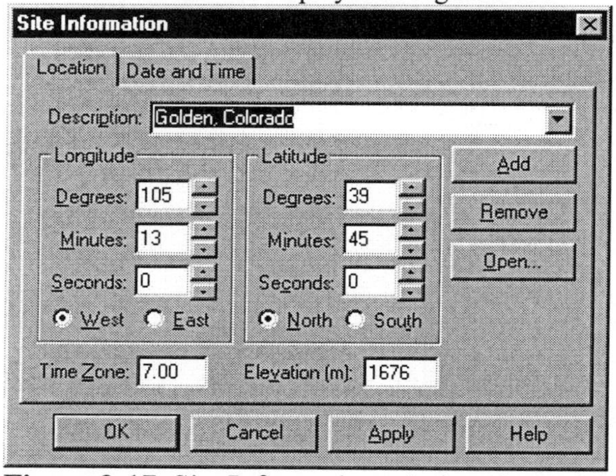

Figure 3-17 Site Information window for Golden, Colorado

5. Click OK (OK). (Do **not** change the date and time.)

6. Find M42 (Look at the previous exercises).

7. Write down its azimuth and altitude from this location.

    Azimuth = _____ °   Altitude = _____ °

8. Compare the values of the altitude and azimuth that you wrote down earlier for M42. They should be the same as those listed in Figure 3-18.

Location	Azimuth	Altitude
Ball State Observatory	132° 41′ 03″	+32° 09′ 09″
Miami, Florida	129° 03′ 58″	+44° 54′ 05″
Golden, Colorado	143° 07′ 21″	+37° 44′ 35″

Figure 3-18 Azimuths and altitudes of M42 from three different locations

9. The reason for the differences in Figure 3-18 is discussed in Chapter 5.

After you add your location to the list of cities, click the Save button 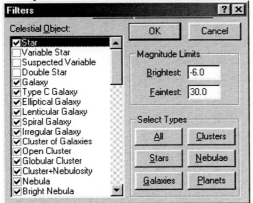 located in the upper left of the sky window on the Standard toolbar. This saves all of the information you just entered into the file called Normal.sky. You can copy this file to a floppy disk or to another folder on your PC, to be used later, in case the information is lost or becomes corrupted.

If you must re-install *TheSky*, the file that the program creates will not have your location or preferences in it. If you wish to copy this file to *TheSky's* folder, it should be renamed (as a backup file) and then copied into the folder (as the file): **C:\Program Files\Software Bisque\ TheSky\User\Documents\Normal.bak.**

It is probably a good idea to keep the number of astronomical objects to a bare minimum until you become proficient in using *TheSky*. To accomplish this, you must first specify what type of astronomical objects you want to display on your computer screen. Exercise F, which follows below, is designed to show how to display some basic celestial information in *TheSky*.

Exercise F: Displaying Celestial Information

1. Click on "Yiew" on the toolbar at the top of the sky window, then go to "Filters." This opens the Filters window that is shown in Figure 3-19.

Figure 3-19 Filters window

This window determines what information you will see on your computer screen. A few changes will probably be necessary to simply things.

2. The first thing you should change is the magnitude limits. Click the button labeled "All" located in the lower right center of the Filters window. This selects all the celestial objects displayed in the sky window.

3. The brightest magnitude should probably be set to about -3.0, while the faintest magnitude to probably +10.0 or less. If you live in a medium-sized city (such as Muncie), then +4.5 is just about right for the faintest magnitude. Magnitudes are discussed in Appendix F.

4. Next, change the information appearing in the "Celestial Object" box. Scroll down the menu of celestial objects and click the Ecliptic, Meridian, and Horizon Lines boxes. These reference lines appear checked in Figure 3-20.

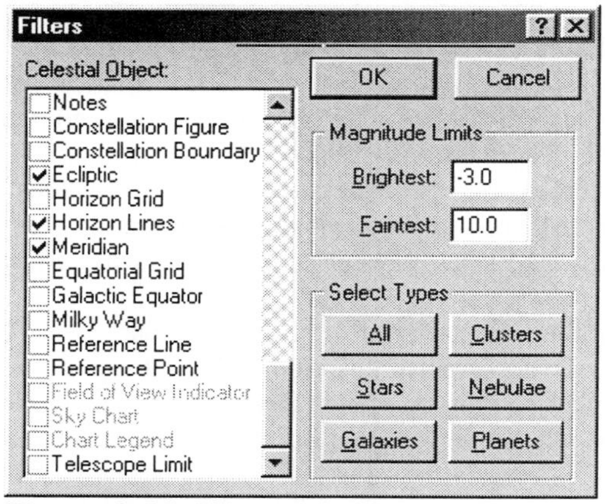

Figure 3-20 Filters window displaying the reference lines

5. Click OK.

6. While you are "looking" South (click ⓢ if necessary), a red dashed line appears perpendicular to the horizon; this is the local celestial meridian. There is also a teal-colored line (teal is blue-green). This line represents the Sun's apparent path in the sky, the *ecliptic*.

7. You can change the colors at any time by using the Preferences option in the <u>V</u>iew menu on the toolbar at the top of the sky window or by clicking the right mouse button anywhere in the sky window. **Appendix E** outlines how to change the Preferences in *TheSky* software.

8. You can change the filters to suit your needs at any time. They are changed the same way as the colors described in the preceding step, Step 7.

Exercise G: Let's Find a Planet

1. Find Uranus in the sky.

2. Click on "<u>D</u>ata" on the toolbar at the top of the sky window and select "<u>S</u>ite Information."

3. Set the location for Ball State Observatory.

4. Click the Date and Time tab.

5. Set the date for October 25, 2000, and the time to 8:00 P.M. These settings are displayed in Figure 3-21.

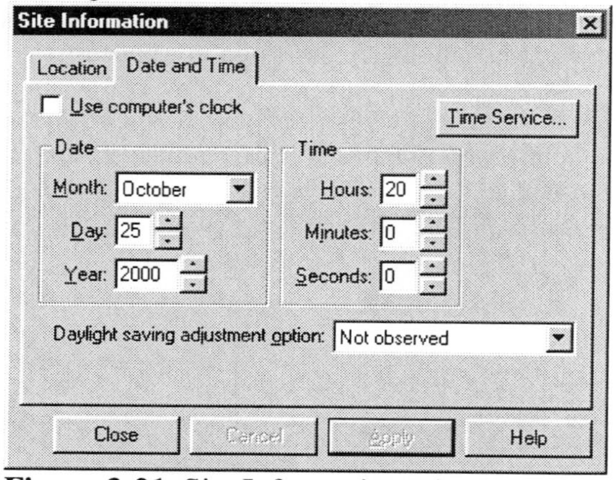

Figure 3-21 Site Information window   showing the date and time.

6. Click Edit on the toolbar at the top of the sky window, then on Find. Click "Planets, Sun, Moon" in the "Common Names" section in the Find window. Now click on Uranus to view known facts about the planet. Be sure to click the More Information button (▼) on the Object Information window.

7. Look at the azimuth and altitude of Uranus. Make a note of these values. **Figure3-22** displays the view of the sky on October 25 at 8:00P.M., facing south from the Ball State Observatory. Uranus is the circled object in the figure.

Figure 3-22 Southern sky from the Ball State Observatory

8. **Figure 3-23** displays the azimuth and altitude information for Uranus.

Object Information

General | Multimedia

Object: Uranus

Type: Uranus   Magnitude: 5.78

Right Ascension: 21h 18m 17s   Declination: -16°27'07"

Azimuth: 185°28'50"   Altitude: +33°10'26"

Uranus
Set: 12:49 AM on 10/25/2000
Rise: 2:36 PM on 10/25/2000
Transit: 7:40 PM on 10/25/2000
RA: 21h 18m 17.1s Dec: -16°27'07"
Azm: 185°28'50" Alt: +33°10'26" with refraction: +33°11'56")
Phase: 99.942%, Apparent magnitude: 5.78
Heliocentric ecliptical coordinates:
l: 319°39'59.2" b: -00°42'14.2" r: 19.951857

Figure 3-23 Object Information window for Uranus

Exercise H: Let's find Uranus from Bombay, India

1. Click "Data" on the toolbar at the top of the sky window.

2. Click on "Site Information." This opens a window like the one displayed in Figure 3-24.

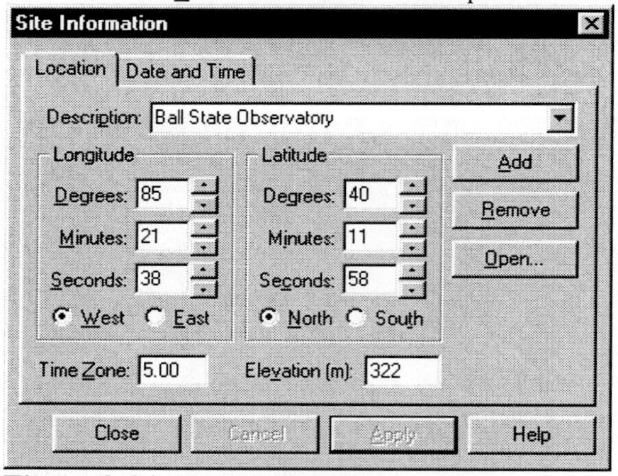

Figure 3-24 Site Information window

3. Make certain the Location tab is selected first.

4. Click on the button labeled Open (Open...).

5. An Open window like the one shown in Figure 3-25 appears.

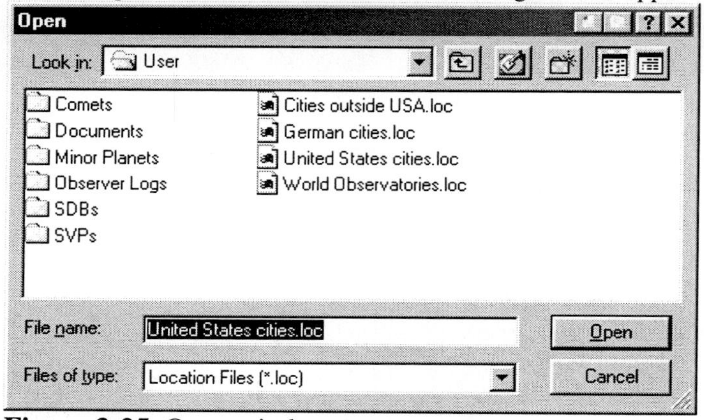

Figure 3-25 Open window

6. Click the file labeled "Cities outside USA.loc".

7. Click "Open."

8. The "Site Information" window reappears.

9. Click the down arrow (▼) located at the end of the description box.

10. Scroll down to Bombay, India as your location. Figure 3-26 displays the Site Information window for Bombay, India.

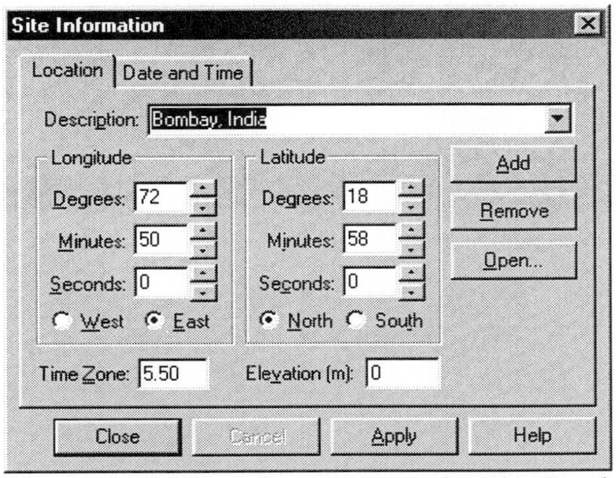

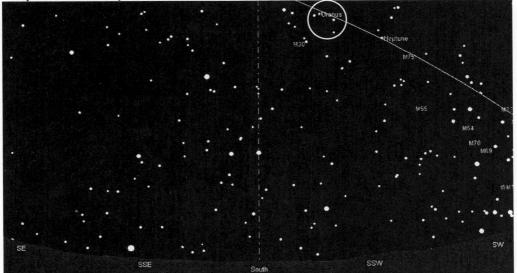

Figure 3-26 Site Information window for Bombay, India

11. Click Apply, then Close.

12. Figure 3-27 displays the southern sky as viewed from Bombay, India. Uranus is the circled object in the figure. Compare this figure to that of Figure 3-22. Note the change in Uranus' position in the sky.

Figure 3-27 Southern sky as viewed from Bombay, India

13. Click on Uranus, and display the Object Information window.

14. Note the altitude and azimuth values of Uranus. The Object Information window displayed looks exactly like the one shown in Figure 3-28.

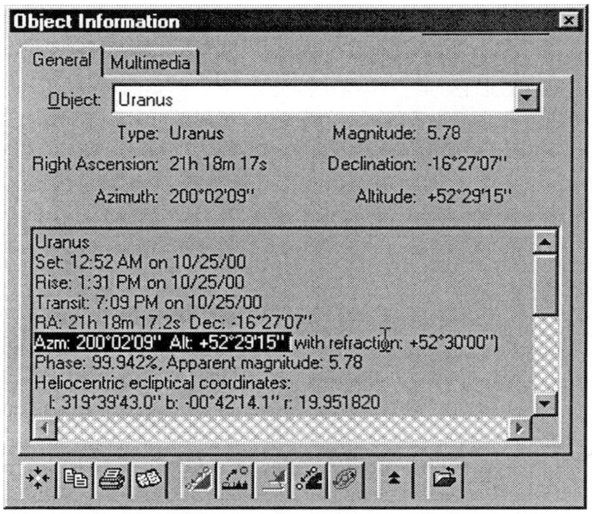

Figure 3-28 Object Information window of Uranus as seen from Bombay

15. Figure 3-29 lists a comparison of the azimuths and altitudes of Uranus as seen from the Ball State Observatory and Bombay, India. Note the difference in the values.

Location	Azimuth	Altitude
Ball State Observatory	185° 28′ 50″	+33° 10′ 26″
Bombay, India	200° 02′ 09″	+52° 29′ 15″

Figure 3-29 Altitude and azimuth of Uranus at two different locations on Earth

Now you know how to set locations and display objects in *TheSky*. It is a simple task if you follow the suggestions given in the previous exercises.

If you have trouble later on, refer back to these exercises. The procedures outlined here are much easier than they seem at a glance. It is best for users to explore different possibilities by changing their location on Earth or *preferences* in the sky window. Remember: the Normal.sky file does not have to be saved, ever!

Once you become proficient in using *TheSky*, then the settings can be changed and saved anytime. In the worst-case scenario, you would have to copy the backup file, Normal.bak to the appropriate folder as Normal.sky or re-install *TheSky*. Have fun and enjoy!

Chapter 4

Naming Objects in *TheSky*

Astronomers need to be able to assimilate and exchange information about specific objects in the sky. Many systems have been devised over the last few thousand years to identify and name the most conspicuous ones. As our technology has advanced, studies of astronomical objects have become more precise. Today it is common for astronomers to assign numerous designations to a single celestial body.

Humans have observed the stars for millennia. Our ancestors named the bright stars as well as larger groups of stars called constellations. As we said in Chapter 1, the ancients named many of the constellations after mythological beasts, gods, demigods, and ordinary household objects. Astronomers continue to use the names of the constellations first recorded by ancient astronomers thousands of years ago. It is here that we may begin to learn about where things are located in the sky and how they are named.

Astronomers officially recognize 88 distinct constellations today. *TheSky* displays all of them quite accurately. From the mid-northern latitudes, you can see over half of them. Most are visible every night from your location at some time during the night. *TheSky* helps you to find them but you must go outside on any clear night throughout the year and look for them yourself.

About a half dozen or so constellations are visible every night from 40°north latitude all year round. These are the circumpolar constellations. They are all located in the northern sky near the North Star, Polaris. Using *TheSky* will definitely help you locate all these constellations easily during any season of the year. Exercise A is designed to help you locate some of the constellations visible from mid-northern latitudes.

Exercise A: View Some Constellations

1. Run *TheSky*.

2. Click Data on the toolbar at the top of the sky window, then Site Information and then click the Location tab.

3. Set your location to 40° north latitude. The city or longitude is not important here! Figure 4-1 displays the appropriate information that should be entered into the Site Information window.

Figure 4-1 Entering 40° north latitude for any observer

4. Click Apply.

5. On the Orientation toolbar, click "North" ().

6. Go to View toolbar and click on the constellation lines (), if they are not already displayed.

7. Click the common names button () on the View Toolbar, if they are not already displayed.

8. Go to the Time Skip toolbar and click the Go Forward () button and watch the motion.

9. List the names of the constellations that do not go below the northern horizon.

_____

_____

_____

_____

_____

_____

_____

These constellations are visible every night throughout the year and are known as *circumpolar constellations*. Exercise B is designed to find a particular constellation in *TheSky*.

Exercise B: Find the Constellation of Orion

1. Click Edit on the toolbar at the top of the sky window, and then Find, or right click the mouse anywhere in the sky window.

2. Click on the "Constellation Labels" in the Common Names window.

3. Use the scroll bar and find the constellation Orion. Figure 4-2 displays the "Find" window.

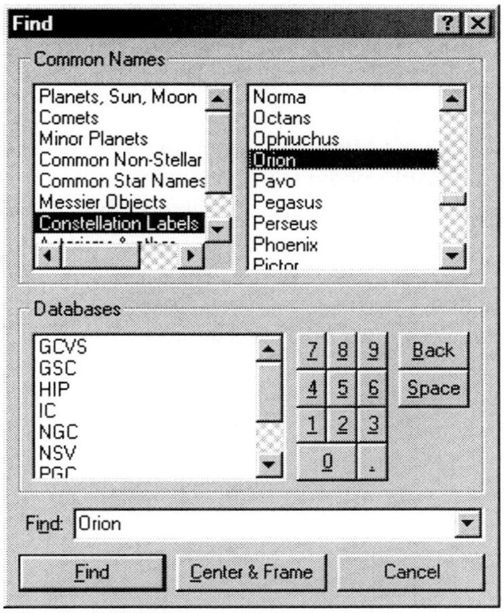

Figure 4-2 Find window

4. Double-click on "Orion."

5. After clicking on "Orion," an Object Information window appears, like the one displayed in Figure 4-3. This box indicates where the constellation is located in the sky and other pertinent information about the constellation, such as the pronunciation of its name.

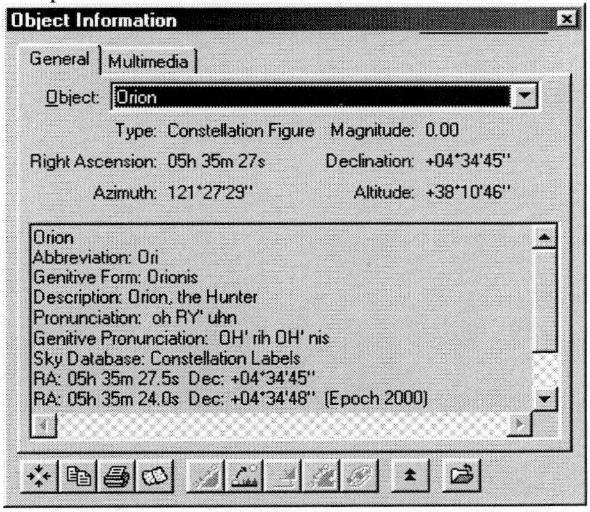

Figure 4-3 Object Information window for Orion

6. Click the Rise, Transit, and Set button ( ) at the bottom of the Object Information window. This displays the time of day that Orion appears to rise, cross your local celestial meridian, and set at your location. The time of meridian transit is the best time to view the object.

> At what time does Orion appear to rise? ____:____
> At what time does Orion appear to set? ____:____
> When is the best time to view Orion? ____:____

7. Next click the Center Object button ( ) at the lower left-hand corner of the Object Information window. This will center Orion in your sky window.

8. Click on some of the objects in the constellation. Did you find M42?

9. If a camera icon is displayed in the sky window, you can view an image of the object by simply clicking on the camera. If the camera icon is *not* displayed in the sky window, then to view an image of the object you must click on the object. After clicking on the object, an Object Information window appears. Now click on the Multimedia tab to access any images of the object. You can display or hide the camera icons in your sky window by choosing either option in the Filters menu.

That's all there is to it. Try finding another constellation—perhaps one you like better than Orion. You can use this exercise to find any astronomical object in *TheSky's* database. Appendix B contains a list of the 88 recognized constellations.

Now let's turn our attention to how astronomical objects are named or designated. Basically, astronomical objects fall into two groups, *stellar* and *nonstellar*. The first group includes *only* the stars. The second group has a more diversified population of astronomical objects. The later group contains objects that lie well beyond our Solar System. The nonstellar group contains objects such as nebulae, star clusters, and other galaxies.

The planets in contrast are not included in either of these groups. Because they are Solar System objects, astronomers usually designate them by their given name.

Let's first discuss how stars are designated. Over the millennia people have used several systems to denote the stars. The numerous cultures that have inhabited our planet have assigned many names to the stars over the years.

Proper Names

Astronomers usually refer to stars as either *prominent* stars or *representative* stars. Prominent stars are the bright stars. Stars, such as those displayed in Figure 4-4, are typically the bright stars seen in the winter months. They are located in the constellations of Taurus, Orion, Canis Major, and Canis Minor.

The dot sizes displayed in the sky window, representing stars, indicate the stars' relative brightnesses, not their true physical sizes. In other words, the bigger the dot, the brighter the star; the smaller the dot, the fainter the star. The magnitude system that astronomers use in observational astronomy is discussed in Appendix F.

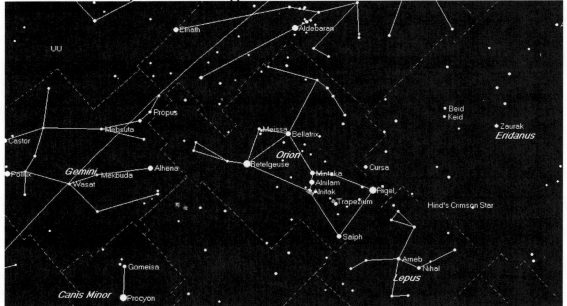

Figure 4-4 Prominent or bright stars of winter

The bright stars of summer are typically those that are displayed in Figure 4-5. These stars are located in the constellations of Lyra, Cygnus, and Aquila.

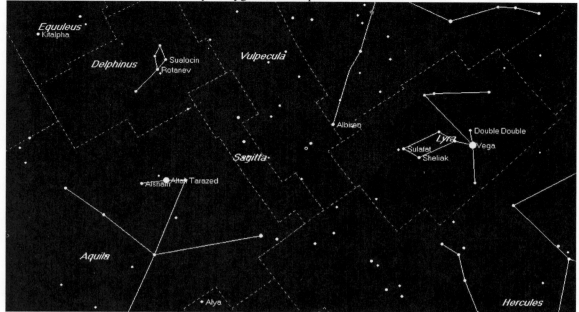

Figure 4-5 Prominent or bright stars of summer

The brightest stars are usually designated with proper names. Only a few dozen stars are frequently referred to by their proper names. Several hundred stars have been named this way, but only a few are easily recognized by name. The bright stars are easily seen after the Sun goes down and twilight falls.

The representative stars, in contrast, are those that are for the most part in the solar neighborhood. They are close by, so to speak. Their distances are within 4 to 8 parsecs of Earth. This is about 13 - 26 light years. Not many of the representative stars have proper names. The representative stars in the Milky Way Galaxy are usually fainter than the Sun's intrinsic brightness. These stars are usually small, cool, and very faint.

On any clear night in spring, you can find the bright star Arcturus in the evening sky. In summer, it might be Antares (heart of the scorpion), or perhaps Vega, the brightest star in the constellation Lyra. In autumn, you might find the star Capella in Auriga, the charioteer. In the winter, it might be Betelgeuse or Rigel in the constellation of Orion. A list of some of the brightest stars is provided in Appendix A. This list includes the stars' designations, proper names, coordinates, apparent brightnesses, spectral types, intrinsic brightnesses, and their approximate distances. Exercises C and D are designed for you to locate some bright stars.

Exercise C: Find the Bright Star Aldebaran

Set your location for Cleveland, Ohio.
Set the date for October 19, 2005.
Set the time to 9:00 P.M.
Set Daylight Savings adjustment option to "North America."
The Site Information window looks like that in Figure 4-6.

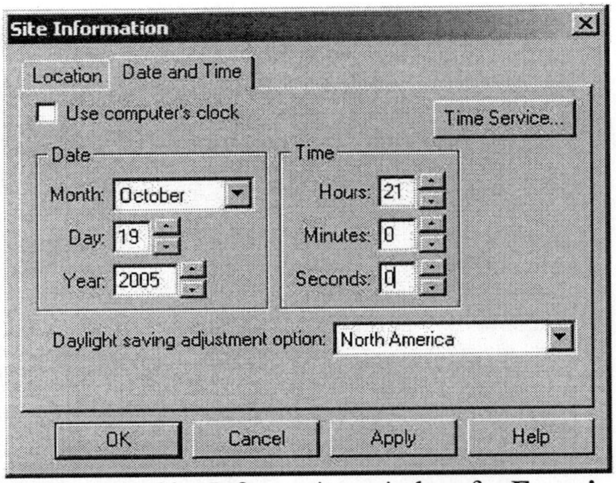

Figure 4-6 Site Information window for **Exercise C**

1. In what constellation is Aldebaran located? _____

2. At what time does Aldebaran appear to rise? ____:____

3. At what time does Aldebaran appear to set? ____:____

4. When is the best time to view Aldebaran? ____:____

Exercise D: Find the Bright Star Altair

Set your location for Golden, Colorado.
Set the date for October 1, 2010.
Set the time to 9:00 P.M.
Set Daylight Savings adjustment option to "North America."

1. In what constellation is Altair located? _____

2. At what time does Altair appear to rise? ____:____

3. At what time does Altair appear to set? ____:____

4. When is the best time to view Altair? ____:____

Bayer Letters

Because it is difficult for most of us to memorize the names of hundreds of stars, Johann Bayer developed a more convenient system in 1603. His system used Greek letters. The stars in the constellations in the northern sky were assigned letters according to their relative brightnesses.

Bayer used the Greek alphabet to designate the stars by their order of brightness in each constellation. Each naked-eye star is labeled with a Greek letter followed by the genitive case (possessive—second person singular) of the Latin name of the constellation in which it is found. This system started with the brightest star in the constellation and went to the faintest.

In 1757 Nicholas Lacaille extended this system to include the southern constellations.

Figure 4-7 shows several stars with their Bayer letter designations displayed.

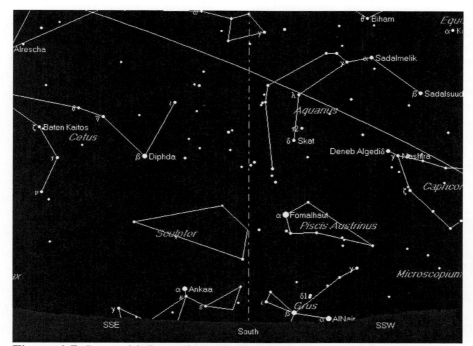

Figure 4-7 Stars with Bayer letter designations

For example, the brightest star in the constellation Lyra is shown in Figure 4-8. Its proper name is ***Vega***. Its Bayer designation is α Lyrae (or Alpha Lyrae). This is like saying "Alpha Lyree." The brightest star in the constellation of Cygnus is Deneb, which is Arabic for "Tail." Its Bayer designation is α Cygni ("Cygnee"). The second brightest star in Cygnus is named Albireo and has a Bayer designation of β Cygni (Beta "Cygnee"). Sometimes constellation names are abbreviated to the first three letters of the constellation name (for example, β Cyg) as in *TheSky*. Appendix B lists the constellation's Latin names, possessive forms and abbreviations of the constellations. The Greek alphabet is provided in Appendix C.

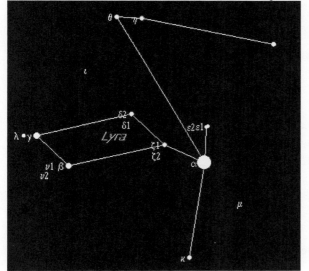

Figure 4-8 Stars in Lyra, the Lyre

After all the Greek letters have been used, the labeling continues with the lowercase letters of the Roman alphabet. If there are enough stars in a particular constellation, then the uppercase Roman alphabet letters are used up to and including the letter ***Q***.

In Latin, some words (nouns and adjectives) are considered to be either masculine or feminine. Masculine words, as a rule, usually end with the letters "us," whereas feminine words end with the letter "a." Masculine constellation names ending in "us" are changed to the second person singular case (genitive) by replacing the letters "us" with the letter "i." Feminine constellation names ending in the letter "a" are changed to the second person singular case by adding the letter "e" to the end of the constellation name.

Note that these are general rules to be followed and by no means are valid in all cases. There are several exceptions. For example, Leo, the Lion is a masculine word. But it doesn't end in the letters "us." The genitive case for this constellation is Leonis. So, Denebola, the second brightest star in Leo, is designated as β Leonis. The same is true for the constellation Virgo, the Virgin. The second personal singular for this constellation is Virginis. So, the brightest star in Virgo is α Virginis. Its proper name is Spica.

And, if this were not confusing enough, a few constellations contain stars that are of about the same brightnesses. In these rare instances, the letters are assigned sequentially as one traces the pattern of the constellation. An example is the constellation Ursa Major (the Great Bear), which is shown in Figure 4-9. Appendix B will help you with the constellation names.

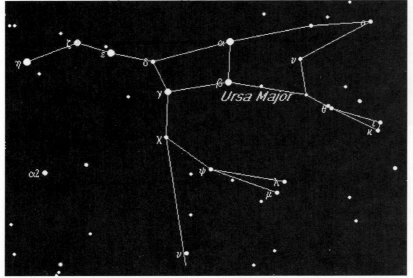

Figure 4-9 Stars in Ursa Major

Exercise E: Find the Bright Star Wasat

Set your location for Atlanta, Georgia.
Set the date for December 19, 2005.
Set the time to 9:00 P.M.
Set Daylight savings adjustment option to "North America."

1. In what constellation is Wasat located? _____

2. What is its Bayer designation? _____

3. What is its magnitude? _____

4. Is Wasat the brightest star in the constellation? _____

5. At what time does Wasat appear to rise? ____:____

6. What time does Wasat appear to set? ____:____

7. What is the best time to view Wasat? ____:____

Exercise F: Find the Brightest Star in the Constellation of Gemini

Set your location for Portland, Maine.
Set the date for January 15, 2020.
Set the time to 9:00 P.M.
Set Daylight Savings adjustment option to "North America."

Find the constellation Gemini.

1. What is the brightest star's proper name? _____

2. What is its Bayer designation? _____

3. What is its magnitude? _____

4. At What time does it appear to rise? ____:____

5. At what time does it appear to set? ____:____

6. When is the best time to view it? ____:____

Flamsteed Numbers

In 1712 astronomer John Flamsteed devised a simple system to designate stars. His system identified many more stars in a given constellation than the letters that Bayer had introduced a century earlier.

Flamsteed assigned numbers to the various stars in each constellation. The major difference in his system is that he ignored stars' apparent brightnesses in the constellation. Each star designated in the constellation is based on its position in the constellation. Figure 4-10 illustrates several constellations and stars with their Flamsteed designations displayed.

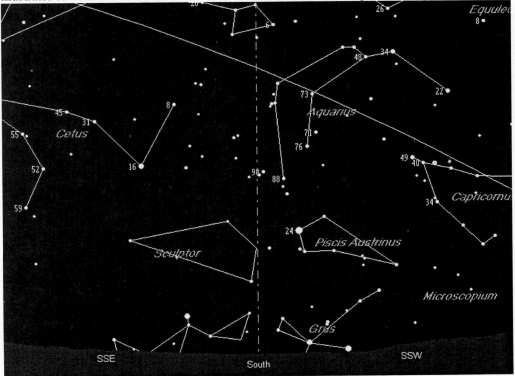

Figure 4-10 Stars with Flamsteed designations

The numbers are assigned in order of their location in the constellation. They are labeled from west to east in each constellation; that is, in the order by which they appear to cross the local celestial meridian. The possessive form of the constellation name follows the number, as in the Bayer system. 6 Cygni is the Flamsteed designation for the second brightest star in Cygnus and is known as β Cygni. Its proper name is Albireo and is displayed in Figure 4-11.

Figure 4-11 6 Cygni, β Cygni, or Albireo

Flamsteed numbers are frequently used to identify stars that are much fainter than the naked-eye stars in each constellation. It is common practice to use the Bayer letters for the brighter stars in the constellations, because they provide some useful information about the apparent brightness of each star. *TheSky* displays both proper names and Flamsteed numbers for the stars visible in the nighttime sky.

Exercise G: Find the Brightest Star in the Constellation of Taurus

Set your location for Chicago, Illinois.
Set the date for February 3, 2005.
Set the time to 9:00 P.M.
Set Daylight savings adjustment option to "North America."

1. What is its proper name? _____

2. What is its Bayer designation? _____

3. What is its Flamsteed number? _____

4. What is its magnitude? _____

5. At what time does it appear to rise? ____:____

6. At what time does it appear to set? ____:____

7. When is the best time to view it? ____:____

Exercise H: Find the Third Brightest Star in the Constellation of Orion

Set your location for Indianapolis, Indiana.
Set the date for December 19, 2005.
Set the time to 9:00 P.M.

Set Daylight savings adjustment option to "Not observed."

1. What is its proper name? _____

2. What is its Bayer designation? _____

3. What is its Flamsteed number? _____

4. What is its magnitude? _____

5. At what time does it appear to rise? ____:____

6. At what time does it appear to set? ____:____

7. When is the best time to view it? ____:____

Binary and Multiple Stars

With the increased use of the telescope by astronomers in the 17th and 18th centuries, another problem arose relating to stellar designation. Astronomers found many individual stars composed of two or more stars that were very close together. They appear as a single star to the naked eye but as two or perhaps three separate stars through the telescope.

Binary stars are most often referred to as "double stars" but this is a misnomer! Astronomers do make a distinction between a double star and a binary star. A double star is two stars that appear very close to each other in the sky. They share the *same direction* in space, but are at *different distances* from us. In other words, they are not gravitationally attached to each other. Binary stars, in contrast, not only share the *same direction* in space but both stars are at the *same distance* from us and orbit each other.

Because it would be a futile task to completely revise all the systems in use today, it is sometimes more suitable to just use the original designation. However, when a star is found to have more than one component, letters are assigned to the various components, in order of their decreasing brightness, after their Bayer designation.

In a binary system, the Roman letter A is assigned to the brighter component (primary component) and the Roman letter B is assigned to the fainter component (secondary component). The possessive form of the constellation name follows the letters, as in the Bayer and Flamsteed designations. If there are more than two stars, then subsequent letters are used (C, D, E, etc.) and the possessive form of the constellation name. There are exceptions to this general rule as well. For example, ζ Ursae Majoris (Mizar) is a well-known binary system. Mizar is the middle star in the handle of the "Big Dipper." The stars in this system are often referred to as Mizar A and Mizar B.

The closest star system to Earth is the α Centauri system (or Alpha Centauri). This star system is actually a tertiary system. The primary component of this system is designated as α Centauri A. The secondary and fainter component is designated as α Centauri B. The faintest component of this system is denoted α Centauri C. The brightest member, A, in this system is very much like our Sun in its size and temperature.

Another famous visual binary system is in the constellation of Cygnus the Swan. It is designated as β Cygni or Albireo. The two stars in this system are separated by 54 seconds of arc and are easily resolved in a small telescope. They are also two stars that have very different surface temperatures and thus very different colors. In a small telescope, they appear as a blue and a yellow star. Figure 4-12 shows the location of Albireo as located in *TheSky*.

Figure 4-12 – Albireo in the constellation Cygnus

Figure 4-13 is a CCD (Charge-Coupled Device) image of Albireo taken at the Ball State Observatory. It is an electronic, digitized image of Albireo taken in three colors. The images were taken with a Photometrics Star 1 camera. Three images were taken through a Red, Green, and Blue filter and later combined to produce the tricolor image displayed in Figure 4-13.

Figure 4-13 CCD image of Albireo taken at Ball State Observatory
(Image provided by Author)

Variable Stars

Another problem, that arose about the same time that astronomers found many stars to be multiple star systems, was that many of these stars appeared to vary in brightness. This change in brightness may be due to a variety of reasons. It may be something intrinsic to the star, such as the very nature of the star itself. Or there may be an extrinsic cause, such as the star being a member of an eclipsing binary system. For whatever reason, the cause of the brightness variation in stars is still designated in a special way.

The last Roman letter used in Bayer's notation was the letter Q. So variable stars begin their designations with the Roman letter R. They are designated by the order of their discovery in the constellation. Thus the first variable star in any constellation is designated with the Roman letter R, then S, T, U, V, W, X, Y, and Z followed by the possessive form of the constellation name. After the letter Z, the letters are doubled RR, RS, RT, to RZ; then SS, ST, SU, to SZ; then TT, TU, TV, to TZ; then UU, UV, UW, to UZ through ZZ and all followed by the possessive form of the constellation name. After the letters ZZ, additional variable stars may be designated with the letters AA, AB, AC, to AZ; then BB, BC, BD, to BZ; then CC, CD, CE, to CZ all the way through QZ each followed by the possessive form of the constellation name in which they are discovered. Note: because there is no letter J in the Roman alphabet, it does not appear in any of the designations. The designations R through QZ take care of the first 334 variable stars discovered in any constellation. If subsequent variable stars are discovered in a constellation, they are designated with a number preceded by the capital letter **V** (for *variable*). Thus, the 335th variable star in any constellation is designated as V335 followed by the possessive form of the constellation name in which it is found. In other words, the 335th variable star discovered in the constellation of Capricornus is designated as V335 Capricorni.

TheSky does not explicitly display variable stars on the desktop window. However, if you are interested in observing variable stars you may contact the American Association of Variable Stars Observers (AAVSO) at 25 Birch Street, Cambridge, Massachusetts 02138-1205. It is a nonprofit scientific and educational international organization of amateur and professional astronomers who study and catalog variable stars. You can also visit their Website at www.aavso.org.

Non-Stellar Objects

Most of the nonstellar objects lie well beyond our Solar System and are a mixed bag of astronomical objects. There are three basic categories of nonstellar objects. They are star clusters, nebulae, and galaxies. Usually these objects are designated with a letter and a number signifying the catalog in which they are found.

Star Clusters

There are two types of star clusters. The type of cluster is indicative of its location in the Milky Way. *Open clusters* are located *in* the spiral arms, and *globular clusters* are *centered on* and *around* the nucleus of our galaxy. The Jewel Box Cluster is an excellent example of an open cluster. It is displayed as the left-hand image in Figure 4-14. Its designation is NGC 4755. Unfortunately, it is not visible from midnorthern latitudes. The second image displayed in Figure 4-14 is one of several hundred globular clusters surrounding the nucleus of the Milky Way. It is displayed as the right-hand image in Figure 4-14 and is designated NGC 1904.

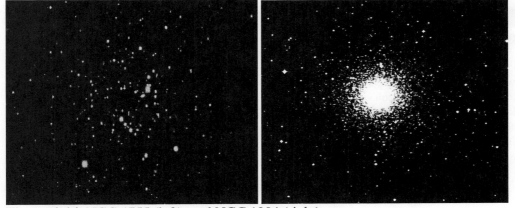

Figure 4-14 NGC 4755 (left), and NGC 1904 (right)

Nebulae

Nebulae are more diversified in nature. They range from objects that are associated with early stages of star formation to those of stellar death.

Dark nebulae are cold regions of gas and dust that loom between the stars in the Milky Way and may harbor *protostars* (infrared stars). This type of nebula is only seen when silhouetted against a bright background of stars or a glowing nebula. One of the best examples of this type of nebula is located in Orion, the Hunter. It is called the Horsehead Nebula, because of its appearance, and is designated as IC 434. An image of it taken from *TheSky's* database is shown in Figure 4-15.

Figure 4-15 Horsehead nebula in Orion, IC 434

Figure 4-16 displays a black-and-white digitized image taken from *TheSky* showing IC 434 and a more extensive view of the region around and near the Horsehead Nebula.

Figure 4-16 Region around and near IC 434

Emission nebulae are associated with newborn stars and sometimes clusters of stars. They glow because the gases in the nebulae are excited by the ultraviolet radiation produced by the surrounding stars. As a result of the ultraviolet excitation, the nebula glows faintly with a pinkish-red color due to the hydrogen in the cloud. A good example of this type of nebula is displayed in Figure 4-17. It is called the Eagle Nebula because of its appearance. This nebula is designated as M16 and NGC 6611.

Figure 4-17 Eagle Nebula, M16, or NGC 6611 in Serpens

Reflection nebulae, in contrast, appear bluish in color. In addition to gas, the cloud contains dust grains that reflect visible light from newly formed stars near the cloud. The Pleiades cluster (Seven Sisters) in the constellation of Taurus is perhaps the best known example of this type of nebulosity. It is a cluster that contains a few dozen stars surrounded by bluish nebulosity. Figure 4-18 shows a picture of the Pleiades taken by Dr. Jordan. It is designation as M 45.

Figure 4-18 Reflection nebula and cluster in Taurus
(Photo courtesy of the author)

Other types of nebulae are associated with the death of elderly stars. When stars like our Sun reach the end of their life cycle, they go through a series of expansions and contractions. Eventually, they shed the outer layers of their atmosphere into space. These nebulae become planetary nebulae.

The most extreme expulsion of stellar material occurs when massive stars reach the end of their lives. They have a tendency to detonate and blow themselves apart. This event causes a sudden brightening of the star, and as a result, a supernova appears in the sky. These nebulae become supernova remnants. The last visible supernova occurred in the Large Magellanic Cloud in 1987. In any case, it is quite a traumatic event for a star.

Planetary nebulae result from stars that were once like our Sun that have gone through their entire life cycle. These nebulae are spherical shells of gas moving outward from the central part of the star. They usually appear as ring-shaped objects with a faint star in the center of the nebula. The planetary nebula in Figure 4-19 is known as the Ring Nebula. It is located in the constellation of Lyra and is usually referred to as M57. It is also designated NGC 6720. The image is a CCD image taken at the Ball State Observatory.

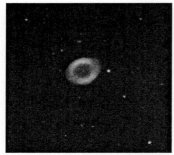

Figure 4-19 Ring nebula in Lyra
(CCD image provided by the author)

Supernova remnants, on the other hand, appear as twisted knots of gas and serve as evidence of violent explosions—to put it mildly. They are more catastrophic than violent! Figure 4-20 displays the supernova remnant in Taurus (the Bull) known as the *Crab Supernova Remnant*. This explosion was observed and recorded by Chinese astronomers in 1054 C.E. It was bright enough, according to historical accounts, to be seen in broad daylight. It is designated as NGC 1952 or as the first object in the Messier Catalogue.

Figure 4-20 Crab Supernova Remnant, M1, in Taurus

Of the types of nebulae discussed here, only the brightest ones are seen either with the naked eye or in binoculars and small telescopes. They appear as smudges, or more often as diffuse faint blurs in a telescope eyepiece. However, they are among some of the most interesting objects in our Galaxy. The brightest nebulae that can be observed and photographed with a small or moderate-size telescope are listed in some of the catalogs that are discussed hereafter.

The Messier Catalogue

In 1781 Charles Messier compiled a list of (cataloged) 110 faint, diffuse, nonstellar objects. He did so to assist those amateur astronomers who were specifically searching for comets. He published the positions of these objects so that other comet hunters would not waste their time making additional observations of them. Messier basically stated that objects in his catalog did *not* move and therefore were *not* comets! Several Messier objects are shown in Figure 4-21.

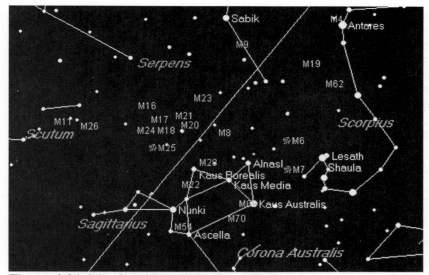

Figure 4-21 Messier objects in and near Sagittarius

The objects in this catalog are numbered from 1 to 110 preceded by the capital letter **M**. A list of the Messier objects is provided in Appendix D.

The New General Catalog

William Herschel made several sky surveys near the end of the 18th century and into the beginning of the 19th century. He used larger and larger telescopes in his observations. Herschel did star gauging or counting measurements of stars in the Milky Way. He also recorded the positions and descriptions of thousands of faint nonstellar objects. In 1888 Herschel's observations were combined with the observations of many other astronomers into what is called the New General Catalogue of Non-Stellar Objects. Several thousand NGC Objects have been compiled for this catalog. The Sombrero Galaxy gets its name from its appearance. It is designated as Object 4594 in the New General Catalog. This particular object is also in Messier's Catalogue and listed as Object 104. With the Object Information window, an image of the Sombrero Galaxy is displayed in Figure 4-22.

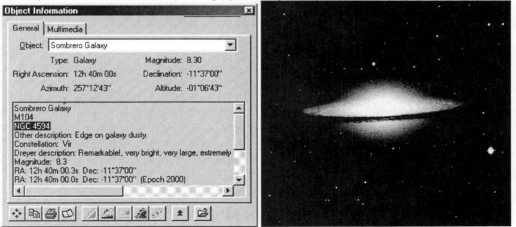

Figure 4-22 Object Information window and NGC 4594, the Sombrero Galaxy

The NGC list of nonstellar objects is too large to include in this workbook. You can easily generate a complete observing list of NGC objects by visiting one of the following Websites: *www.ngcic.org* or *www.ngic.org/oblstgen.htm*. It is possible to create a complete list of NGC objects in any constellation at this Website.

Later two supplemental catalogs were published. They contained information on more recently discovered objects (other than NGC objects). They are called the First and the Second Index Catalogues, respectively. The objects contained in both of these catalogs are denoted with a number preceded by the letters "IC." The astronomical objects listed in the IC catalogs are, as a rule, fainter than many of those listed in the NGC catalog. The IC lists contain several thousand more nonstellar objects. You can also create an observer's list of the IC objects in any constellation by accessing the same website as for the NGC objects. With the Object Information window, an image of the IC 5152 is displayed in Figure 4-23.

Figure 4-23 Object Information window and IC 5152

Hubble Space Telescope

Since its launch in April 1990, the Hubble Space Telescope has been quietly orbiting Earth and returning remarkable images and data on many astronomical objects. It has compiled and assimilated more information in the last decade than all earth-based optical telescopes have historically, combined! The Object Information window is displayed for the star GSC 5847:2333 in Figure 4-24. This star has the proper name Diphda and also has the designation of β or 16 Ceti.

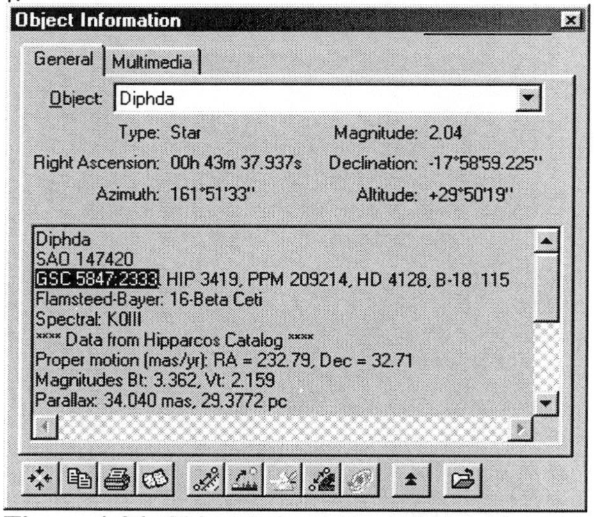

Figure 4-24 Object Information window for Object GSC 5847:2333, Diphda

Hubble has compiled data on millions of stars and non-stellar objects as well. The information about these stars is stored on about 100+ CD-ROMs. Each star is cataloged with a number preceded by the letters **GSC,** for Guide Star Catalog.

Other Catalogs

Other catalogs that list information about stars are used in *TheSky*. The Harvard Smithsonian Astrophysical Observatory published a catalog that is often used by astronomers. The Observatory compiled data on about 250,000 stars. Stars in this catalog are labeled with a number and preceded by the letters SAO. Figure 4-25 displays the Object Information window for the star SAO 147420. This is still the star β Ceti.

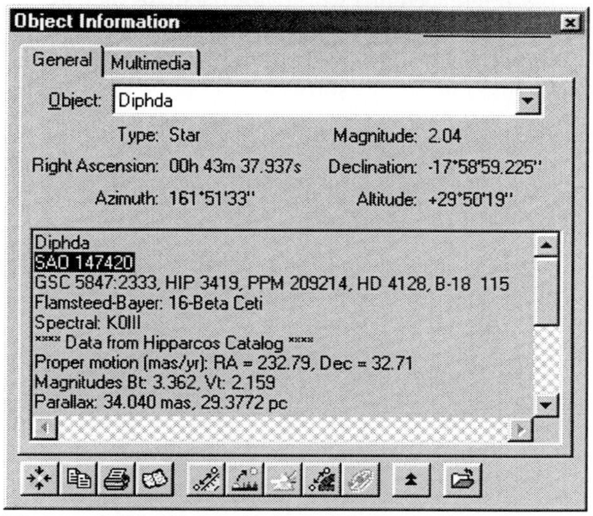

Figure 4-25 Object Information window for star SAO 147420, β Ceti, Diphda

Another catalog is the Henry Draper Catalog. This catalog, compiled between 1918 and 1924, contains spectral information on over 225,000 stars. Stars in this catalog are designated with a number preceded by the letters HD. The Object Information window for the star HD 4128 is displayed in Figure 4-26. Notice that this star is still β Ceti.

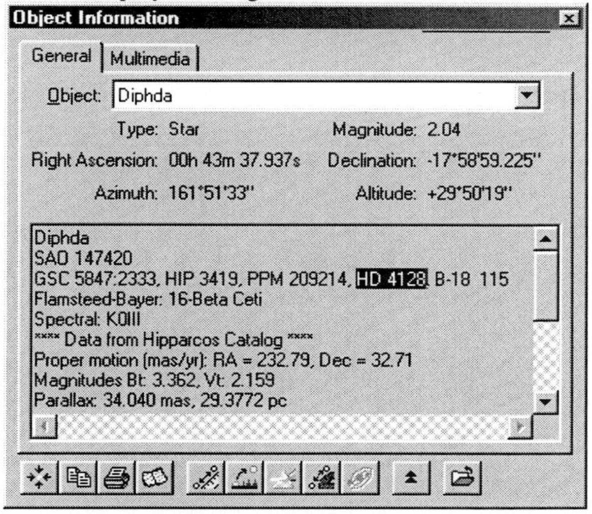

Figure 4-26 Object Information window for star HD 4128, Diphda, β Ceti

There are two other catalogs available in *TheSky* database. They are the PPM and HIPPARCOS catalogs. PPM is an acronym that stands for the Positions and Proper Motion catalog. The stars in this catalog are designated with a number preceded by the letters "PPM." Diphda is designated as Object PPM 209214. HIPPARCOS is an acronym that stands for "High Precision Parallax Collecting Satellite." The stars in this catalog are designated with a number preceded by the letters "HIP." Diphda is designated in the HIPPARCOS catalog as

Object HIP 3419. The distances listed, for many of stars in *TheSky*, have been derived from the data collected by the HIPPARCOS satellite.

Today we know of many diverse types of astronomical objects, and their designations can be confusing. Only computers have the ability and capacity to store and assimilate information about these objects. Every day more and more statistics are compiled and stored in databases. This is one reason why using *TheSky* is both convenient and resourceful.

Numerous catalogs are accessible to those of us interested in observational astronomy. Each is compiled for a specific reason or to detail specific information about certain groups of celestial objects. This information has been cross-referenced in a worldwide database such as the one at the Centre de Donnees Astronomiques de Strasbourg. Astronomers around the world can access this database via the World Wide Web each day (cdsweb.u-strasbg.fr/CDS.html). Most of the designations previously discussed are used in *TheSky* software.

Chapter 4

TheSky Review Exercises

Use *TheSky* to complete the following exercises. The first one has been done for you.

Before starting *TheSky*, insert the CD-ROM into your drive. If *TheSky* has already been started, close it, then insert the CD-ROM, and restart *TheSky*.

TheSky Exercise 1: Let's Find Object NGC 7000

1. Set Date to ***October 25, 2006***.
   Set Time to 8:00 P.M.
   Make sure that the "Daylight saving adjustment <u>o</u>ption" is set to "Not observed."

   Set Location to Ball State Observatory
   Longitude: 85°21' 38" <u>W</u>est   Latitude: 40° 11' 58" <u>N</u>orth
   Set Time Zone to +5.0.
   Set Elevation to 322 meters.

   Note: If you wish to save this location for future exercises, the properties must be change in the United States cities.loc file in the folder C:\ProgramFiles\Software Bisque\TheSky\ User (see Chapter 1 Figures 1-23 and 1-24 or Chapter 2 Exercise A).
   When finished, the settings should look like those displayed in Figure 4-27.

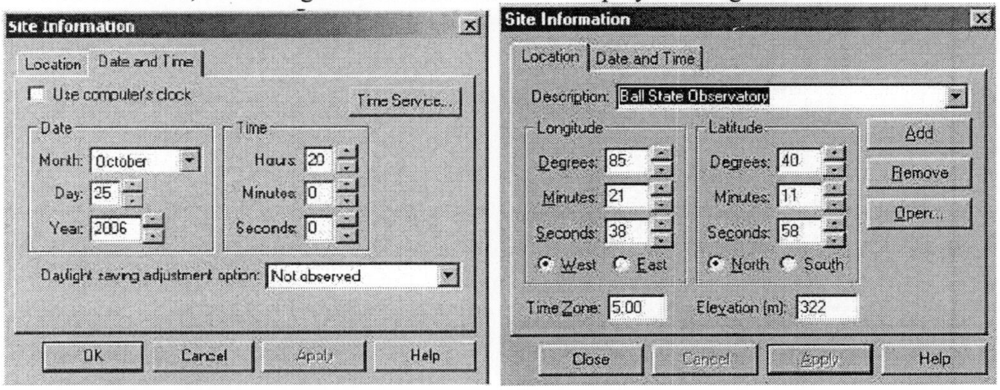

Figure 4-27 Changing the date and time

2. Click <u>E</u>dit on the toolbar at the top of the sky window, then click *<u>F</u>ind*.

3. A shortcut to this command is to just type the letter "**F**." Or, right click mouse anywhere in the sky window. Either way, the Find window appears, like the one shown in Figure 4-28.

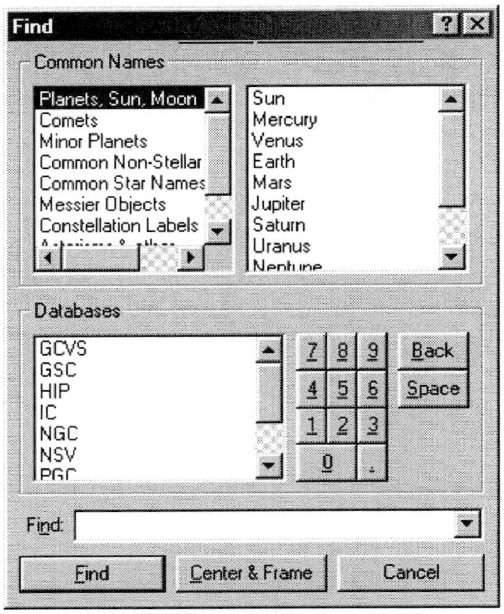

Figure 4-28 Find window

4. Let's find an NGC object, say NGC 7000.

5. In the Databases box of the Object Information window, click "NGC," and then click the number 7000 on the keypad provided at the right. The Databases box will then look like Figure 4-29.

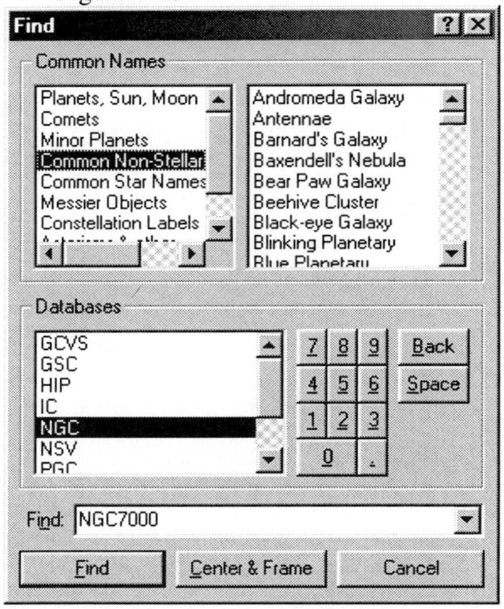

Figure 4-29 Finding NGC 7000

6. Click the Find button below the Databases index.

7. After clicking Find, an Object Information window opens like the one that is shown in Figure 4-30.

Figure 4-30 Object Information window for NGC 7000

7. Click the More Information button (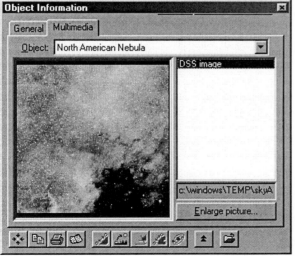) in the Object Information window and data in regard to NGC 7000 will appear. This is displayed in **Figure 4-31**.

Figure 4-31 Information on Object NGC 7000

8. Use the scroll bar to look at the data provided on NGC 7000.

9. Click the ⊹ button to center NGC 7000 on your desktop.

10. Now click the Multimedia tab. When your CD is inserted into your PC's CD-ROM drive, an image of NGC 7000 is displayed like the one shown in **Figure 4-32**.

Figure 4-32 Image of the North American Nebula, NGC 7000

Now, it's your turn!

Complete the following exercises using *TheSky*.

If you wish, leave the location set to the Ball State Observatory. Otherwise, set it to your location. The important thing is *not* to change the date and time.

TheSky Exercise 2: Find Object M30

1. What is the NGC number associated with M30? _____

2. In what constellation is M30 located? _____

3. What is its apparent magnitude? _____

4. What type of non-stellar object is this? _____

5. Does it have a common name? _____ If so, what is it? _____

6. Click the Multimedia tab to view an image of M30.

TheSky Exercise 3: Find Uranus

1. What time does Uranus appear to rise? ____:____ Transit? ____:____ Set? ____:____

2. What is the apparent magnitude and phase of Uranus? _____

3. Click the Multimedia tab in the Object Information window to view images of Uranus. Now click and open the file named "Uranus.txt" at the bottom of the list and answer the following questions.

   a) What is the orbital period of Uranus (in Earth years)? _____

   b) What is the rotational period of Uranus (in Earth days)? _____

   c) What is the mass of Uranus (in Earth masses)? _____

   d) The atmosphere of Uranus is made up primarily of what gas? _____

   e) What is the surface gravity on Uranus? _____
      (G is in Earth's gravity)

   f) What *special* characteristics does Uranus have? _____
      _____

TheSky Exercise 4: Find Jupiter

1. What time does Jupiter appear to rise? ____:____ Set? ____:____

2. When does Jupiter appear to cross the local celestial meridian? _____

3. What is the apparent magnitude and phase of Jupiter? _____

4. Click the Multimedia tab in the Object Information window and view images of Jupiter. Now click and open the file named "Jupiter.txt" at the bottom of the list.

a) What is the orbital period of Jupiter (in Earth years)? _____

b) What is the rotational period of Jupiter (in Earth days)? _____

c) What is the mass of Jupiter (in Earth masses)? _____

d) The atmosphere of Jupiter is made up primarily of what gas? _____

e) What is the surface gravity on Jupiter? _____

f) What *special* characteristics does Jupiter have? _____
_____

TheSky Exercise 5: Find Object M42

1. What time does M42 appear to rise? ____:____

2. What is its apparent magnitude? _____

3. What is its NGC number? _____

4. In what constellation is it located? _____

5. What type of object is it? _____

6. Click the Multimedia tab to view a few pictures of this object.

TheSky Exercise 6: Find Object NGC 224

1. What time does NGC 224 appear to rise? ____:____

2. What is its apparent magnitude? _____

3. What Messier number does it have? _____

4. In what constellation is it located? _____

5. What type of object is it? _____

6. Does it have a common name? _____

7. Click the Multimedia tab to view a few pictures of this object.

TheSky Exercise 7: Find Object IC 446

1. What time does IC 446 appear to rise? _____:_____

2. When does this object transit your local meridian? _____:_____

3. In what constellation is it located? _____

4. What type of object is it? _____

5. What is its apparent magnitude? _____

6. Click the Multimedia tab to view a picture of this object.

TheSky Exercise 8: Find Object NGC 2244

1. What time does NGC 2244 appear to rise? _____:_____

2. When does this object transit your local meridian? _____:_____

3. In what constellation is it located? _____

4. What type of object is it? _____

5. What is its apparent magnitude? _____

6. Click the Multimedia tab to view a picture of this object.

TheSky Exercise 9: Find Object M79

1. What time does M79 appear to rise? _____:_____

2. When does this object transit your local meridian? _____:_____

3. In what constellation is it located? _____

4. Does it have an NGC number? _____ If so, what is it? _____

5. What type of object is it? _____

6. What is its apparent magnitude? _____

7. Click the Multimedia tab to view a picture of this object.

TheSky Exercise 10: Find Object M13

1. What time does M13 appear to rise? _____:_____

2. When does this object transit your local meridian? _____:_____

3. In what constellation is it located? _____

4. Does it have an NGC number? _____ If so, what is it? _____

5. What type of object is it? _____

6. What is its apparent magnitude? _____

7. What is the object's angular size? _____

8. Click the Multimedia tab to see a picture of this object.

TheSky Exercise 11: Locate and Identify an Object

1. On the toolbar at the top of the sky window click on <u>O</u>rientation, then click on <u>M</u>ove To, or right click the mouse button and click on <u>M</u>ove To. Now enter the following information into the Move To dialog box:

   Right Ascension: $13^H 30^M 00^S$ Declination: $48° 00' 00''$ North
   Epoch: 2000

2. What object is located at the following coordinates? Right Ascension $= 13^H 29^M 53^S$ and Declination $= +47° 11' 54''$ _____

3. What is its Messier Number? _____

4. Does it have a Common Name? _____

5. What is its NGC Number? _____

6. What time does this object appear to rise? _____:_____

7. When does this object transit your local meridian? _____:_____

8. In what constellation is it located? _____

9. What type of object is it? _____

10. What is its apparent magnitude? _____

11. What is the object's angular size? _____

12. Click the Multimedia tab to see a picture of this object.

TheSky Exercise 12: Locate and Identify an Object

1. Set date to July 4, 2017.
   Set time to 9:30 P.M.
   Make sure that the "Daylight saving adjustment <u>o</u>ption" is set to "North America."

   On the toolbar at the top of the sky window click on <u>O</u>rientation, then click on <u>M</u>ove To and enter the following information into the Move To dialog box:

   Right Ascension: $20^H 45^M 5.0^S$ Declination: $40° 15' 15.0''$ North
   Epoch: Current

   Now click on the <u>O</u>rientation, then click on Zoom <u>T</u>o and then on <u>B</u>inocular 50°.

2. What constellation is located in the center of the field of view? _____

3. Are there any Messier Objects in this constellation? _____

4. List some of them: _____, _____, _____, _____, _____, _____, _____, _____.

5. Do any of them have Common Names? _____

6. What are the coordinates of M 29?

    R.A. = ____H ____M _____S   Declination = ___ _____° _____′ _____″

7. What time does this object appear to rise? _____:_____

8. When does this object transit your local meridian? _____:_____

9. What type of object is it? _____

10. What is its apparent magnitude? _____

11. Does it have an NGC Number? _____. If so, what is it? _____

12. Click the Multimedia tab to see a picture of this object.

Chapter 4

TheSky Review Questions

Answer the following review questions:

1. In what order are the Bayer letters assigned to stars? _____

2. Is δ-Capricorni brighter than α-Capricorni? _____
   How do you know? _____

3. In what order were Flamsteed numbers assigned to stars? _____

4. How many nonstellar objects did Messier include in his catalogue? _____

5. Why did Messier compile his list, and how were these objects selected?
   _____

   _____

6. What nomenclature is used to identify variable stars? _____

7. Is V 143 Cyg a valid variable star designation? _____ If not, why not?
   _____. What is its correct designation?
   _____.

8. Is V469 Lyrae a valid variable star designation? _____

9. If a star is found to be multiple, how are its various components designated?

   _____

10. What is the order of discovery of TZ Orionis? _____

Chapter 5

Locating Celestial Objects in *TheSky*

Have you ever searched for a friend's house in a different city and found it next to impossible to find? Did you get lost? Perhaps there were only a few street signs, or perhaps that your friend didn't give you very clear instructions to follow. You probably stopped at a gas station and asked for directions or went to a telephone and called your friend to ask for more detailed directions.

Even though street signs may offer little assistance, think of what it would have been like to not have had any signs at all! We continually encounter problems in finding our way around or in getting from one place to another when it involves directions. In observational astronomy, general procedures have been adopted to help avoid these problematic situations. In other words, astronomers have developed a system of coordinates to locate the positions of celestial objects very precisely.

Angles and Coordinate Systems

This chapter contains exercises that help you learn how to use coordinate systems in observational astronomy. To understand an astronomer's method of locating objects in the sky, it helps to have a working knowledge of angles and a three-dimensional view of the sky.

Coordinate systems are usually set up in such a way that we must first choose a *frame of reference*, that is, a place from which our detailed measurements of the sky are made. In this instance, the Earth is the *only* frame of reference we have! We begin by envisioning the sky above us, as ancient astronomers did, to be a large sphere. In doing so, this will amount to determining just two angles to locate objects on this sphere.

Many people have difficulty measuring angles. To avoid the most common mistakes made, we will take a very simplistic approach. We begin our discussion of angles by reviewing how angles are measured and what units are used to express angular measurement.

When it comes to angles, astronomers view things in the sky in two ways. The first is *angular separation*, or *how far apart* two objects appear to be in the sky. The second is *angular size*, or *how large* objects appear to be in the sky. Either way, angles are used to describe the separation and size of astronomical objects.

Angles are measured in units called degrees (°). There are 360 of them in a full circle. Therefore, one degree represents 1/360th of a full circle. Degrees are usually used for making large angular measurements. For example, there are two stars in the Ursa Major known as the *pole pointers* (α and β Ursae Majoris) because they always point to Polaris. They have an *angular separation* of about 5°. The stars α and β Ursae Majoris, the pole pointers, in the Ursa Major as seen in *TheSky* are illustrated in **Figure 5-1**.

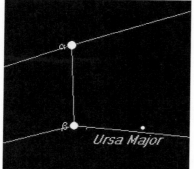

Figure 5-1 α and β Ursae Majoris, Pole Pointers in Ursa Major

The Sun and (full) Moon, in contrast, have an angular size of about ½°. **Figure 5-2** displays actual images of the full Moon and Sun taken from *TheSky's* image database. Even though the Sun and full Moon have about the same angular size, the Sun is over 400 times larger

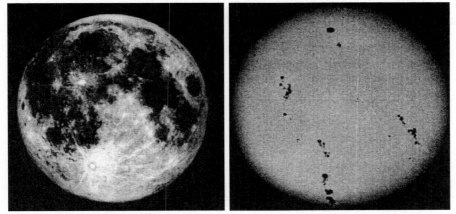

Figure 5-2 Moon and Sun

than the Moon and over 400 times farther away than the Moon is from Earth. Despite this, the Sun appears to be about the same size as that of the full Moon in Earth's sky.

To measure an angle smaller than a degree, you need additional units. Therefore, 1° can be further divided into 60 equal smaller divisions called *minutes of arc* or *arcminutes*. Arcminutes are usually designated with an apostrophe mark ('). Don't confuse this mark with linear measurements of feet!

An average human eye (with 20/20 vision) can resolve two objects that have an angular separation of about 12 arcminutes. This happens to be the angular separation between the two stars that make up the "middle star" in the handle of the *Big Dipper*. Their proper names are Alcor and Mizar. They are also designated as 80 and ζ Ursae Majoris, respectively. They are shown in **Figure 5-3**. Try to find Mizar and Alcor some clear night. They are easy to locate in the sky, and you can consider them an eye test.

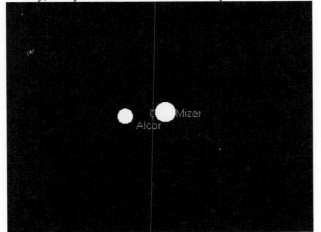

Figure 5-3 Mizar and Alcor: ζ and 80 Ursae Majoris, respectively

If you look at Mizar and Alcor through a small telescope, which has a greater resolution than the eye, you notice immediately that the angular separation between Alcor and Mizar is much greater and that Mizar has a very close companion. It is, indeed, a binary star system.

To measure the angular separation between ζ Ursae Majoris A and B (also known as Mizar A and B), you need smaller angular units. To accommodate very small angles, such as between binary stars, we further divide 1 arcminute into 60 equal smaller divisions called *seconds of arc* or *arcseconds*. Arcseconds are usually designated with a quotation mark ("). Don't confuse this mark with linear measurements of inches! An arcsecond represents the smallest angular unit used in measuring angles. There are 1,296,000 of them in a full circle. ζUrsae Majoris A and B are separated by only 14 arcseconds in angle. To see both stars easily you need, at least, a pair of binoculars or a small telescope.

A realistic example of how small an arcsecond is would be for a friend to hold a penny between their finger and thumb. Now look at the penny that they are holding from a distance of 2½ miles away! This is how large an arcsecond appears in the sky. The angle equivalencies are as follows:

$$360° \text{ (degrees)} = \text{full circle}$$

$$1° \text{ (degree)} = 60' \text{ (arcminutes)}$$

$$1' \text{ (arcminute)} = 60'' \text{ (arcseconds}$$

This is the most common procedure that people use to measure angles in astronomy.

However, another system in observational astronomy uses *time units* to measure angle instead of the conventional degrees, minutes, and seconds of arc. This system is based on a rotating Earth. Because Earth rotates once on its axis (full circle) every 24 hours, it is convenient to divide the sky into 24 equal parts instead of 360. This system employs a sphere divided into 24 equal intervals called *hours*. Each 1-hour interval represents 15° in angle. These hour intervals can be further divided into *60 equal* smaller intervals called *minutes*, and each minute interval can be further divided into *60 equal* smaller intervals called *seconds*. This is the system that you will eventually learn and the one that astronomers use to locate celestial objects in the sky.

It is convenient to use equivalents to convert angular units to time units and vice versa. The following equivalents are very helpful:

$$15° = 1^H$$
$$15' = 1^M$$
$$15'' = 1^S$$

Other equivalent expressions are used in the conversion of angular units to time units, too. Because $1^H = 60^M$, which equals 15° of angle, then

$$4^M = 1°$$

and, because $1^M = 60^S$, which equals 15' of angle, then

$$4^S = 1'$$

You are now about ready to find things in the sky. First, however, some knowledge of coordinate systems is helpful. There are two coordinate systems used in the sky. Before discussing them, it is essential to review the coordinate system used on Earth.

The Geographic Coordinate System

The coordinate system that is most familiar to everyone is the system that is used on Earth's surface. It is the *geographic coordinate system*. In the geographic coordinate system, the surface of Earth is the frame of reference, and the *geographic equator* is the basic plane of origin from which and along which the coordinates of *latitude* and *longitude* are measured.

Not only are angular measurements used in this system but *directions* are also specified. In the geographic coordinate system, the *geographic equator* is a fundamental great circle. A *fundamental great circle* is a circle whose plane passes through the center of a sphere and has a diameter equal to that of the sphere. It divides the sphere into two equal halves. **Figure 5-4** illustrates the location of the north and south geographic poles as well as the geographic equator. By virtue of the way it is drawn, the geographic equator divides Earth into two equal halves, north and south.

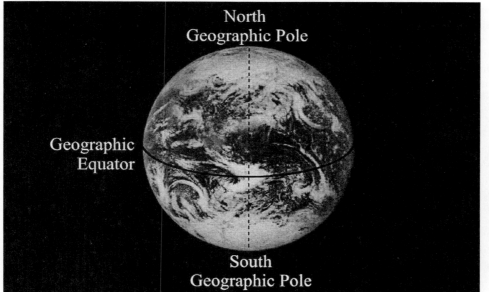

Figure 5-4 Geographic Equator

    The north and south geographic poles are the two points about which Earth rotates and therefore defines its axis of rotation. The geographic equator is actually defined as the circle drawn *exactly* halfway between (90° from) the north and south geographic poles.

    As stated earlier, the geographic equator is a great circle and the primary reference circle in the geographic coordinate system. Circles that are formed by intersections of the sphere with planes not passing through the center of the sphere have diameters smaller than that of the sphere and are consequently called *small circles*. So circles parallel to the geographic equator are defined as small circles (or latitude circles).

    Secondary reference circles are used to divide Earth in half in an east–west direction. These secondary reference circles are also great circles and are shown in **Figure 5-5**.

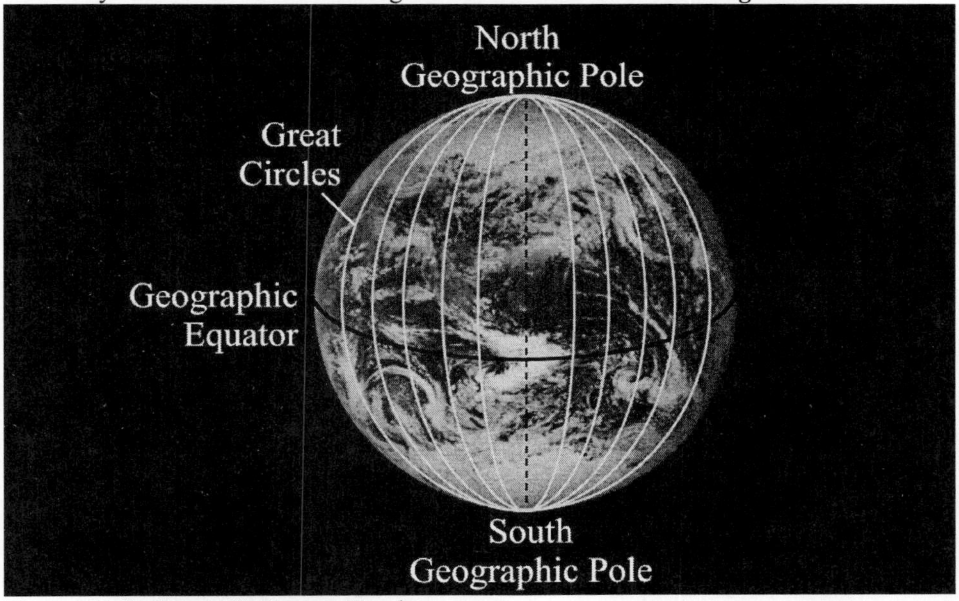

Figure 5-5 Great Circles on Earth

These great circles are drawn on the Earth in a north–south direction through the north and south geographic poles. The great circle passing through Greenwich, England (agreed on internationally since 1884) is called the *reference meridian* or the *prime meridian*. All other great circles that pass through the poles and are parallel to the prime meridian are also known as *meridians* (or longitude circles) on Earth.

To specify locations on the surface of Earth, geographers use the coordinates of latitude and longitude. *Latitude* is an angle that is measured north or south of the geographic equator as viewed from Earth's center, along a *meridian* or *longitude circle*. It is measured from the geographic equator to the latitude (small) circle on which the location is situated. The measurement of the latitude of Muncie, Indiana, is displayed in **Figure 5-6**.

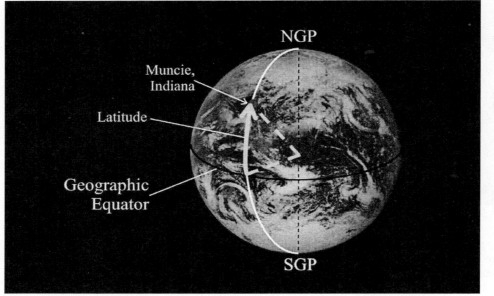

Figure 5-6 Latitude of Muncie, Indiana

Muncie, Indiana is located 40°11′15″north of the geographic equator. The *values* of latitude range from 0° to 90° (0°at equator and 90°at the NGP or SGP). Directions are specified as being either north or south of the geographic equator.

Longitude is an angle measured around Earth's equator, as seen from its center. The *values* of longitude range from 0° to 180°. Directions are designated as being either east or west of Greenwich, England (prime meridian). The measurement of the longitude of Muncie, Indiana, is displayed in **Figure 5-7**.

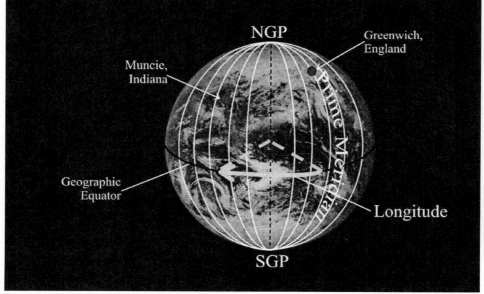

Figure 5-7 Longitude of Muncie, Indiana

Muncie, Indiana is located 85° 21′ 15″ west of Greenwich, England. The geographic coordinate system is an easy system to use to find the location of any city, observatory, or missile silo on Earth. The location of the Ball State Observatory is 40° 11′ 58″ N and 85° 21′ 38″ W.

In *TheSky* you can change these coordinates to any location on Earth. Usually the latitude and longitude of the place where the observer lives is used. *TheSky* lets you move to any location on the surface of Earth (0°–90° N–S in latitude and 0°–180° E–W in longitude).

The Student Version of *TheSky* defaults to sites in the United States. To observe from other sites around the world, you must access the <u>S</u>ite Information window and change the location. To do this, click <u>D</u>ata on the toolbar at the top of the sky window then click on <u>S</u>ite Information and then the "Location" tab and finally click the "<u>O</u>pen" button. A list of sites will be displayed in the Open window menu. Only location files have the extension "*loc*" after them. **Figure 5-8** and **Figure 5-9** display the Site Information window with the Date and Time tab and the Location tab displayed, respectively.

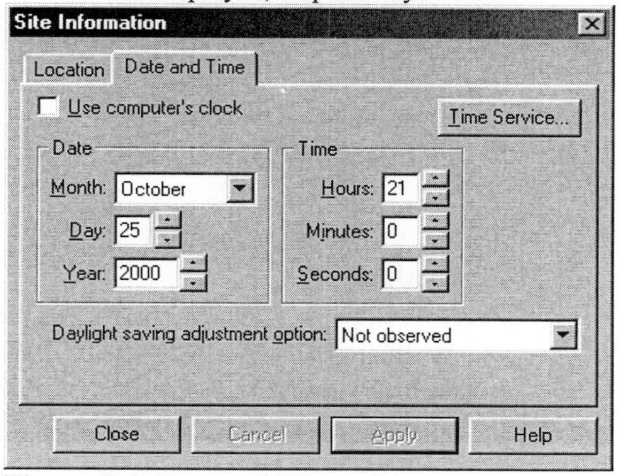

Figure 5-8 Date and Time tab

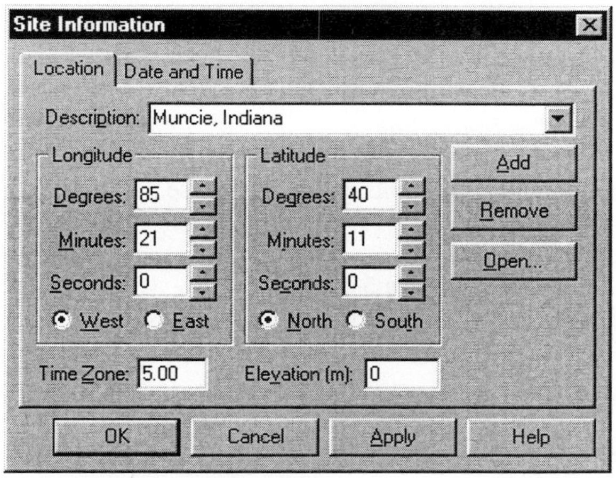

Figure 5-9 Location Tab

Figure 5-10 displays the location files in the Open window menu. After you have clicked on any of the location files, then click the "<u>O</u>pen" button. This lets you access any of *TheSky's* location databases. Once you have opened a location database, the Site Information window reappears. You must then click the Description line to display the contents of that database. After clicking the Description line, scroll down the list to choose the appropriate site. You must then click <u>A</u>pply then Close.

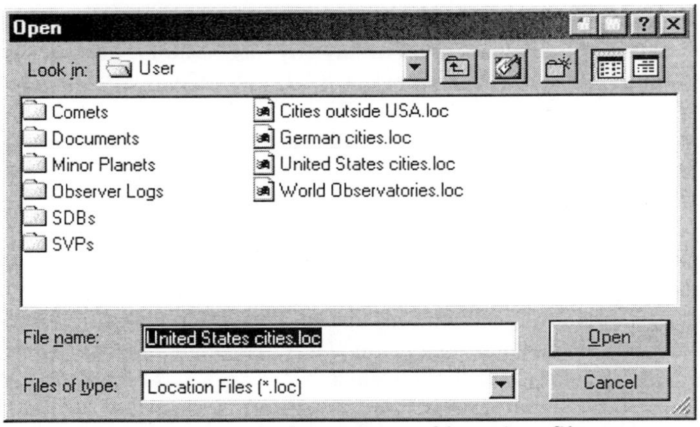

Figure 5-10 Open window menu of location files

If your desired location is not found in the list of indexed sites, then you must type it in and add it to the list. The site's latitude and longitude must be known before adding it to *TheSky's* database. **Exercises A** and **B** illustrates how you can view the sky from the north geographic pole and the geographic equator.

Exercise A: View *TheSky* From the North Geographic Pole

1. Run *TheSky*.

2. Go to <u>D</u>ata on the toolbar at the top of the sky window.

3. Select <u>S</u>ite Information.

4. Select the <u>L</u>ocation tab.

5. Highlight and delete the current selection on the Descri<u>p</u>tion line.

6. Type the words North Geographic Pole in the description line.

7. Set your latitude to 90°N. The longitude is unimportant in this exercise because all the longitudes converge at the poles.

8. The Site Information window should look like that in **Figure 5-11**.

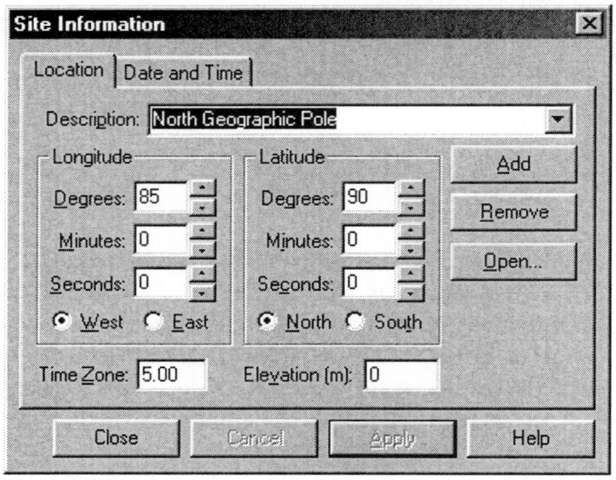

Figure 5-11 Setting your location to the north geographical pole

9. If you click the zenith button (), you should be able to notice that the bright star Polaris in the constellation of Ursa Minor is directly overhead as seen in **Figure 5-12**.

Figure 5-12 Zenith as seen from the north geographic pole

10. The view displayed in **Figure 5-12** is changed back to the Normal view. To accomplish this, go to Orientation on the Standard Toolbar and click on it. You must then click on "Zoom To" and choose "Naked Eye 100°." This resets the sky back to the Normal view window.

11. Enter a time skip of 10 minutes (⟦10 minutes⟧).

12. Push the "Go Forward" time step button (⟦▶▶⟧).

13. Stop the time skip by pushing the Stop button (⟦■⟧).

14. Describe the motion observed. _____ _____.

15. Do any objects go below the horizon? _____

16. Click the East button (⟦E⟧) on the Orientation Toolbar. What does the motion look like near the eastern horizon? _____

17. Push the Reset button (⟦▲⟧) to repeat this exercise or another direction button and observe the motion.

18. Close the file and do not save it, or continue with **Exercise B**.

Exercise B: Viewing *TheSky* From the Geographic Equator

1. Go to Data on the toolbar at the top of the sky window.

2. Select Site Information.

3. Select the Location tab.

4. Highlight and delete the current selection on the Description line.

5. Type the words Geographic Equator in the description line.

6. Set your latitude to 0°N. To see Polaris, set you latitude to 2°N.

7. Set your longitude for 0°W.

8. Click the North button ( ) to look North.

9. Enter a time skip of 10 minutes (10 minutes).

10. Push the Go Forward time step button ( ).

11. Stop the time skip by pushing .

12. Describe what happened. _____

13. Do any objects go below the horizon? _____

14. Click the East button ( ) to look East.

15. Describe the motion. _____

16. Push the Reset button ( ) to repeat this exercise or click another direction button.

17. Close the file and do not save it.

The Celestial Sphere

Having reviewed Earth coordinates, now you can apply this knowledge to the sky. For the astronomer, the sky now becomes the frame of reference. Since ancient times the sky has been known as the *celestial sphere*. The celestial sphere is an "apparent sphere" with its radius centered on the observer standing on Earth and extending outward to infinity.

If the Earth's axis of rotation is extended outward to infinity, it appears to touch the sky at two points. It is around these two points that the celestial sphere *appears* to rotate. These two points are defined as the *north* and *south celestial poles*. Their location in the sky is directly above the north and south geographic poles. If a line is drawn *exactly* halfway between (90° from) the north and south celestial poles, it defines a fundamental great circle in the sky known as the *celestial equator*. The celestial equator lies directly above the geographic equator in the sky. And, like the geographic equator on Earth, divides the sky into two equal halves, north and south. The celestial sphere is illustrated in **Figure 5-13**.

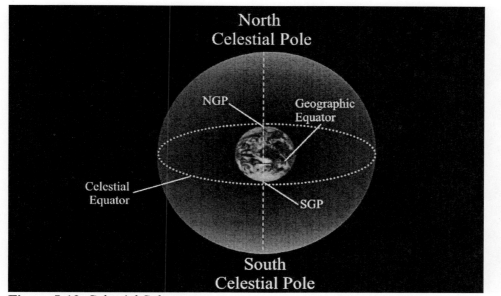

Figure 5-13 Celestial Sphere

It is on this sphere that astronomers put their coordinates to locate celestial objects. There are two coordinate systems that are used in observational astronomy.

The Horizon Coordinate System

The easiest sky coordinate to use in observational astronomy is based on the position of the observer, and uses the horizon as the fundamental great circle. This system, which is similar in design to the geographic coordinate system, is based on the astronomical horizon and is known as the horizon coordinate system.

The astronomical horizon is determined as a result of the direction of gravity for an observer's position on Earth. It is a plane that is perpendicular (90°) to the direction of gravity, and coincides with the observer's line of sight. That is, where land and sky appear to meet. This plane extends outward to infinity and divides the celestial sphere into two equal halves again; that is, half the sky is above the horizon and half the sky is below it.

Before setting up this coordinate system, two points in the sky need to be defined. These points are first located in order to draw the fundamental great circle that is used in this system. The first point to locate is defined as the *zenith point*. It is the point in the sky that is *directly above* your head. The second point to locate is known as the *nadir*. It is a point in the sky that is exactly 180° from your zenith. That is, it is located at the point in the sky directly below the observer–on the other side of the Earth. These two points in the sky represent the "poles" in this system.

The great circle that is used as the fundamental reference circle in this coordinate system is the astronomical horizon. It is then defined as a circle drawn on the celestial sphere that is *exactly* halfway between (90° from) the zenith and the nadir. **Figure 5-14** illustrates the location of the zenith, nadir, and astronomical horizon projected onto the celestial sphere.

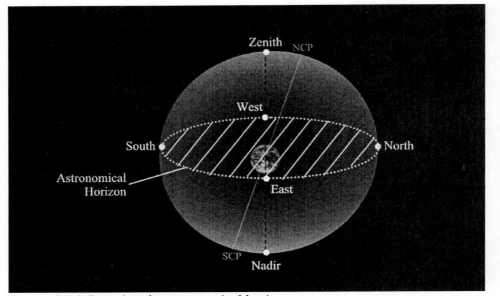

Figure 5-14 Locating the astronomical horizon

Using the horizon coordinate system to find celestial objects once again involves measuring just two angles. The first, is an angular measurement made from the astronomical horizon along a secondary great circle known as a *vertical circle* (remember: all circles passing through poles are great circles) to the object. The object's position is somewhere on this vertical circle with respect to the astronomical horizon.

The first angle is measured *above* the astronomical horizon and is defined as an object's *altitude*. The values for altitude range from 0° to 90°. The altitude of any object on the horizon is 0° and that at the zenith is 90°. The measurement of altitude on the celestial sphere is depicted in **Figure 5-15**.

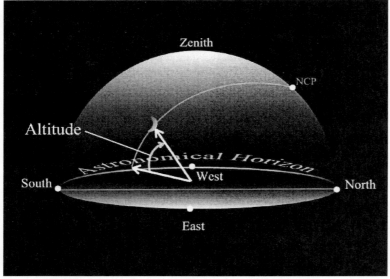

Figure 5-15 - Measuring altitude on the celestial sphere

The second angle is measured *around* the astronomical horizon. Like longitude on Earth, astronomers measure this angle from a well-defined point on the astronomical horizon. This well-defined point is the *north point* on the horizon. It is the reference point from which to measure this angle. The angle is defined as an object's *azimuth*.

Azimuth is the angle measured from the north point on and along the astronomical horizon toward east, and clockwise about the zenith point, to the vertical circle passing through the object. The values for azimuth range from 0° to 360°.

The major compass points on the horizon are commonly referred to as the *cardinal points* (N, E, S, and W). Their azimuths are 0°, 90°, 180°, and 270°, respectively. The North Point actually has two values, 0° and 360°. Therefore, the azimuth of any object that appears to be rising due east is 90°. The measurement of azimuth on the celestial sphere is illustrated in **Figure 5-16**. Once you have mastered these coordinates, this is an easy and useful coordinate system for locating celestial objects from any locale on Earth.

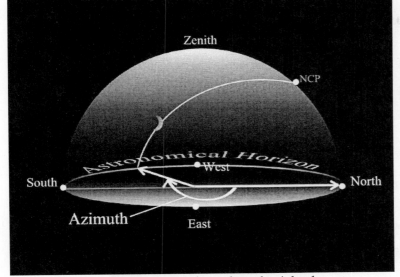

Figure 5-16 Measuring azimuth on the celestial sphere

In *TheSky* there is a *horizontal grid* that displays approximate altitudes and azimuths of celestial objects. It is a red sphere with grid marks ( ), located on the View toolbar. When this button is clicked, the horizontal grid is activated. Red lines appear in the sky window and represent altitude and azimuth in the sky. The angular separation between each line of altitude is 10° of angle, and that of azimuth is 15° of angle.

When you are looking north, Polaris is located above the north point on the horizon. For any observer at 40° N latitude, Polaris is located about four (4) red lines above the northern horizon, or at an altitude of 40°, as shown in **Figure 5-17**. Polaris is the circled object in the figure.

Figure 5-17 Altitude of Polaris for an observer at 40° N latitude on Earth

Displaying horizon coordinates for any object in the sky is a simple task. Click on any object desired, and its coordinates are displayed in an Object Information window. For example, when you left-click on Polaris with the mouse, an Object Information window appears like the one displayed in **Figure 5-18**.

Figure 5-18 Object Information window for Polaris

The Object Information window displays the proper name of the object. Clicking on the red sphere in the Object Information window usually displays an object's horizon coordinates. However, because of a programming glitch the blue sphere must be clicked on instead. The

way to avoid confusion is to click the More Information button () located at the far right on the toolbar in the Object Information window.

The Object Information window is displayed again in **Figure 5-19**. This figure illustrates what the Object Information window looks like after clicking the More Information button. It displays information about the object that has been chosen. Scrolling down the menu bar one click, you find the Bayer and Flamsteed designations for Polaris. By choosing the More Information button, you avoid any problem involving the program glitch.

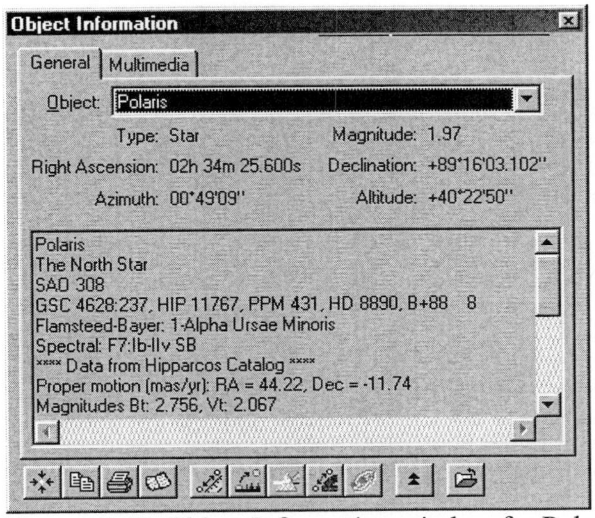

Figure 5-19 Object Information window for Polaris

In any event, any object's altitude and azimuth is displayed for any time of day, any day of the year, from any location on Earth. The horizon coordinates are displayed in **Figure 5-20** for the star Polaris.

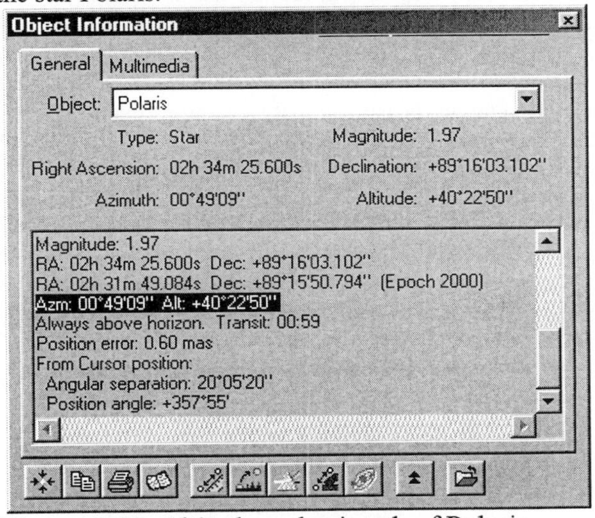

Figure 5-20 Altitude and azimuth of Polaris

There are two major disadvantages of the horizon coordinate system. As Earth rotates on its axis, objects in the sky appear to move westward 15 degrees every hour. Consequently, their altitudes and azimuths constantly change with time. Second, because Earth is spherical your location on Earth makes objects appear to be in different places in the sky.

The problem is most noticeable if you observe a star or constellation from two locations that are separated by several hundred miles along a north–south direction on Earth. A case in point: Remember that when you viewed Polaris from 40° N latitude, its altitude above the northern horizon was about 40°. When you viewed Polaris from the north geographic pole (90°N), like you did in **Exercise A**, it was located at your zenith, an altitude of 90°. The fact that the altitudes of stars and constellations change when viewed from different locations on Earth led ancient astronomers thousands of years ago to conclude that the Earth was round.

Observing objects in the sky works best with the horizon coordinate system if you and a friend with whom you might want to share this information live in the same town. It is a simple system to learn, and convenient to use. *TheSky* provides this information so that anyone can go outside and locate celestial objects easily. The main thing to remember is the shortcomings of this system. That is, the horizon coordinates *constantly change with time* and *depend on your location*.

Most amateur astronomers can see several hundred celestial objects without the aid of a telescope. With a telescope, however, thousands–even millions–of additional objects come into view. Of these, the stars are the most numerous. Because distances to stars are enormous, they are considered infinitely far away. Once you understand this assumption, you can mark their positions with coordinates on the celestial sphere.

The Equatorial Coordinate System

You might be wondering by now if there is a coordinate system that remains fixed in time and is independent of an observer's location on Earth. Would such a system be difficult to devise because of a rotating Earth? In fact, there is such a system, and it is really quite simple to use.

The frame of reference is once again the celestial sphere. It is most appropriate for this system to be fixed to it. The fundamental great circle in this system, however, is the celestial equator from which this coordinate system derives its name. It is known as the *equatorial coordinate system*. The equatorial coordinate system measures angles from and around the great circle that lies halfway between the projection of Earth's geographic poles onto the celestial sphere. The celestial equator is precisely halfway between (90° from) these two points in the sky and is often considered as a projection of Earth's equator onto the sky. These projected poles (north and south celestial poles) are the two points in the sky that are "fixed" and about which the entire celestial sphere appears to rotate. This coordinate system is affixed to the celestial sphere and moves with it.

At first it may seem that finding one's way around the sky using this system is at best difficult. But, soon you realize that we once again end up measuring just two angles that are *exactly* like the ones in the geographic coordinate system used on Earth.

The first angle is measured from the celestial equator along a vertical circle perpendicular to (90°) the celestial equator that passes through an object in the sky. Vertical circles on the celestial sphere are also known as *hour circles*. The reason, of course, is that the sky has been divided into 24 equal parts along the celestial equator, with each part representing 15° of angle or 1 hour of time.

The angular measurement is also directional. That is, it is measured both north and south of the celestial equator. This angle is defined as an object's *declination*. It is measured exactly like latitude is on Earth.

However, so that this coordinate is not confused with the geographic coordinate of latitude the notation of $\pm$ is used instead of north and south. The +sign denotes angles measured north of the celestial equator while the –sign denotes angles measured south of the celestial equator. The measurement of declination on the celestial sphere is illustrated in **Figure 5-21**.

As stated earlier, the celestial equator is a great circle and the primary reference circle in the equatorial coordinate system. Circles that are formed by intersections of a sphere with planes not passing through the center of the sphere, have diameters smaller than that of the celestial sphere and are consequently called *small circles*. So, the circles that are parallel to the celestial equator are defined as small circles. These small circles, above or below the celestial equator, are known as *declination circles*.

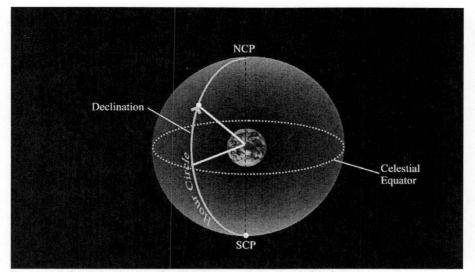

Figure 5-21 Measuring declination on the celestial sphere

As in other coordinate systems, secondary reference circles are used to divide the sky in half in an east–west direction. These secondary reference circles are also great circles. The celestial equator is the fundamental great circle along which the second coordinate is measured. This second coordinate requires a reference point on the celestial equator from which to measure this angle—a Greenwich, England, so to speak!

Any point on the celestial equator could be chosen as the point of origin for the second coordinate. However, the point that is chosen on the celestial equator is the point where the celestial equator and the *ecliptic* intersect. The intersection point is near the constellation boundaries of *Pisces* and *Aquarius*.

The *ecliptic* is defined as the apparent path of the Sun. Over the course of a year the Sun *appears* to move eastward along this path. It appears to move through 13 constellations. That's right, 13! The 13 constellations through which the Sun appears to move each year are known as the *zodiacal constellations* or simply, the *zodiac*. In antiquity, and in *astrology* today, only 12 constellations are recognized as constellations of the zodiac. The Sun's apparent motion along the ecliptic is actually cause by Earth's orbital motion, or its revolution around the Sun.

At two locations on the celestial sphere, the celestial equator and ecliptic intersect each other. These are defined as the *vernal* and *autumnal equinoxes*. Astronomers have chosen the vernal equinox as the reference point or more importantly as the zero point for the secondary fundamental plane, which *is* the vertical circle passing through this point and the point 180° from it. It so happens that the vernal equinox is also the point on the celestial sphere where the Sun appears to cross the celestial equator on or about March 21 each year. The vertical circle passing through the two equinoxes is known as the *equinoctial colure*.

An angle is then measured eastward along the celestial equator from the vernal equinox to an hour circle passing through any object on the celestial sphere. This angle is known as the coordinate of *right ascension*. Right ascension is the *only* angle in observational astronomy that is measured in hours, minutes, and seconds of time rather than degrees, minutes, and seconds of arc (angle). Its value ranges from 0^H to 24^H, with subdivisions of minutes and seconds. The measurement of right ascension on the celestial sphere is illustrated in **Figure 5-22**.

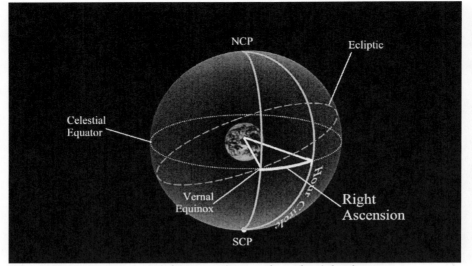

Figure 5-22 Measuring right ascension on the celestial sphere

At first this system may seem confusing and impossible to use. Why use hours instead of degrees? It becomes clearer when you think about this angle being based on a rotating Earth. As Earth rotates on its axis, the *celestial sphere appears* to rotate counterclockwise about the north celestial pole. That is, if an observer goes outside and faces north, after a while you will notice that the sky's motion about the NCP (Polaris) is counterclockwise. Looking south, in contrast, you notice that the motion of the sky is from east to west. Perhaps **Exercise C** will help you visualize these motions.

Exercise C: Viewing the Motion About the North Celestial Pole

1. Run *TheSky*.

2. The location doesn't matter.

3. Click the North button ( ) on the Orientation toolbar.

4. Set the time skip to 3 minutes on the Time Skip toolbar (3 minutes) and press the "Enter" key.

5. Click the Go Forward button ( ).

6. Notice that the motion about the NCP (Polaris) is counterclockwise.

7. Click the "Stop" button ( ).

8. Now, click the South button ( ).

9. Click the Go Forward button ( ) again.

10. Notice that the motion now is from east to west.

11. Click on the Equatorial Grid (blue sphere) button (⊕). Notice too that the blue lines representing right ascension and declination also move from east to west.

12. Close the file and do not save it.

As a result of Earth's rotation, each vertical circle passing through the celestial poles appears to move 15° westward in the sky as seen in **Exercise C**. The measurement of this motion is with respect to an imaginary reference circle drawn through your zenith.

The reference circle, passing through your zenith, is a north–south line drawn from the south point on the horizon to the north point on the horizon. This north–south line is defined as your *local celestial meridian*. The local celestial meridian in the previous exercise is found in the middle of the sky window while facing due south. In **Exercise C** you should notice that *one* of these blue vertical circles crosses your local celestial meridian each hour. In *TheSky*, the default setting does not display the local celestial meridian. However, **Exercise D** demonstrates how you can display it.

Right ascension may be thought of as a measurement of celestial longitude of astronomical objects. It is very similar to longitude measured on Earth. The basic difference between geographic longitude and right ascension is that *longitude* on Earth *is* an angle *measured* both *east and west* of Greenwich, England, whereas *right ascension* is an angle *measured only eastward* from the vernal equinox. Both angles, of course, are measured around fundamental great circles (equators).

TheSky displays equatorial coordinates for all the objects in its data files. Right ascension and declination of the cursor's position is found near the bottom center in the main window of *TheSky*. These coordinates are *always* displayed here. Whenever the right ascension and declination of anything in the sky is desired, just click on it. An Object Information window opens and displays the information about the object.

In *TheSky* there is an Equatorial Grid that displays approximate right ascensions and declinations of celestial objects. It is a *blue sphere* with grid marks (⊕) located on the View toolbar. When this button is clicked, the equatorial grid is activated. Blue lines appear in the sky window and represent right ascension and declination in the sky. The angular separation between each line of declination is 10° of angle, and that of right ascension is 1 hour of angle.

Displaying equatorial coordinates for any object in the sky is a simple task. Click on any object desired and its coordinates are displayed in an Object Information window as the horizon coordinates were earlier. Clicking on the blue sphere in the Object Information window usually displays an object's equatorial coordinates. However, due to a programming glitch the red sphere must be clicked on instead. This confusion is also avoided by clicking the More Information button (⤓) located at the far right on the toolbar in the Object Information window as was done for the horizon coordinates. In fact, the Object Information window should be displayed in this fashion from hereon!

In any event, the object's right ascension and declination are displayed for any time of day, any day of the year, from any location on Earth. The equatorial coordinates for Polaris are displayed in **Figure 5-23** after clicking the More Information button. Unlike horizon coordinates, right ascension and declination of celestial objects are the *same* for every observer on Earth. They are independent of time and your location.

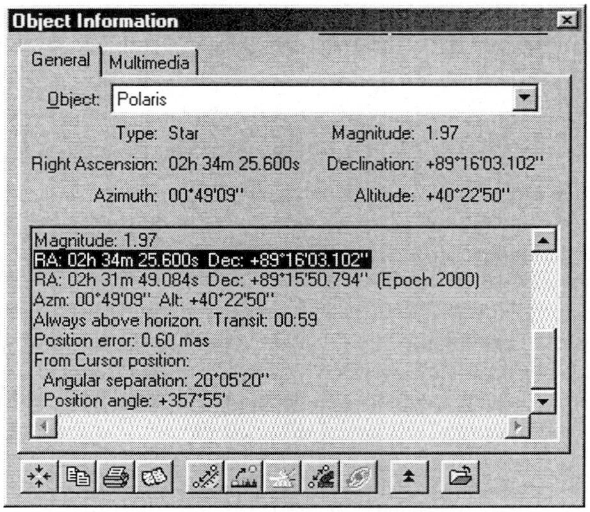

Figure 5-23 Right ascension and declination of Polaris

At this point, it would be helpful for you to visualize the orientation of the sky. Imagine standing outside and facing North as was done in **Exercise C**. Looking up into the sky, draw a line (in your mind's eye) from the north point on the horizon, through your zenith, to the south point on the horizon. This is your *local celestial meridian*. In *TheSky*, this reference line is not displayed in the default Normal.sky window. However, it is an easy task for you to display this reference line. **Exercise D** illustrates how you can display this reference line and the orientation of the celestial sphere.

Exercise D: Orientation of the Celestial Sphere

1. Run *TheSky*.

2. Open the file Normal.sky. Click View on the toolbar at the top of the sky window.

3. Click on "Reference Lines." This displays a window with several options. The Reference Lines menu is displayed in **Figure 5-24**.

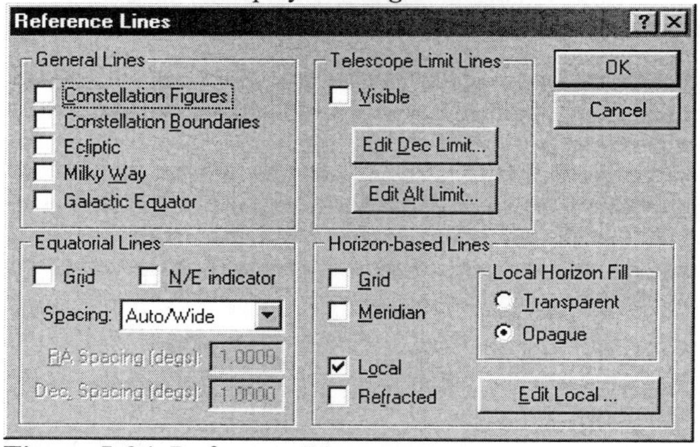

Figure 5-24 Reference Lines menu

4. In the Horizon-based Lines section, click on the Meridian option. This option displays the local celestial meridian in the Normal.sky window. **Figure 5-25** illustrates the *Reference Lines* window after the changes are made.

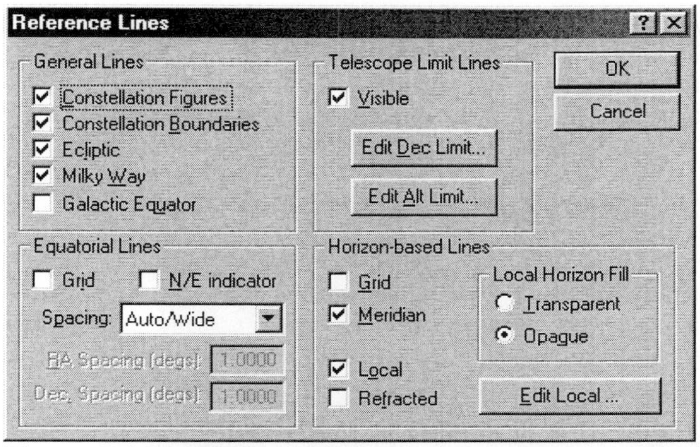

Figure 5-25 Reference Lines window in *TheSky* after changing the options

5. On exiting *TheSky*, the program prompts you, asking, "Save changes to Normal.sky?" Click <u>Y</u>es! This action saves the Normal.sky file with the settings just completed.

6. The Save changes to Normal.sky window is illustrated in **Figure 5-26.** Click <u>Y</u>es!

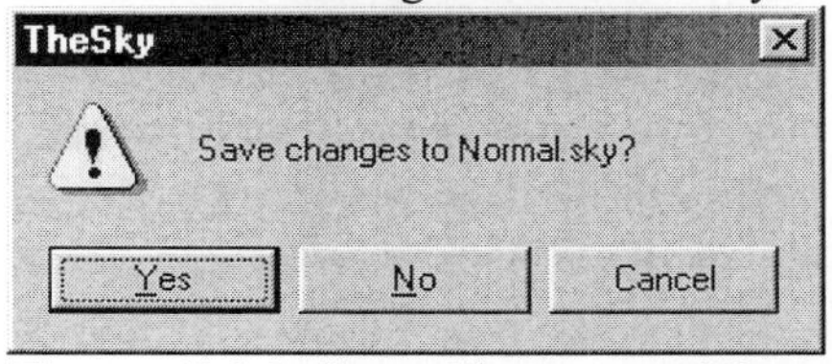

Figure 5-26 Save Changes to Normal.sky window

7. After clicking the <u>M</u>eridian option, *TheSky* displays it in the Normal.sky window. The default color setting in *TheSky* is a *vertical red-dotted line.* **Figure 5-27** displays the local celestial meridian looking north. You can change the color of this line at any time. You can find the instructions for changing the colors and other Preferences in *TheSky* in **Appendix E**.

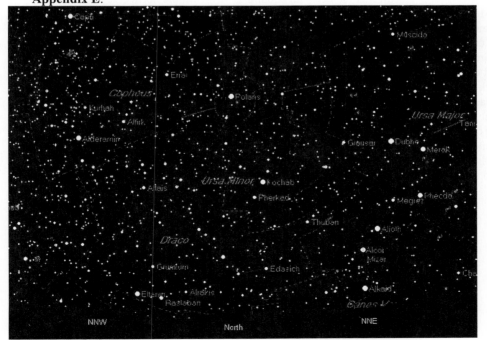

Figure 5-27 Local celestial meridian facing north at 40° N latitude

While facing north, you can easily locate Polaris or more correctly the north celestial pole. They are located approximately 40° above the north point on the horizon for an observer at 40°N latitude as shown in **Figure 5-27**. The north celestial pole is situated on the local celestial meridian, whereas Polaris is not! In **Figure 5-28** *TheSky* displays the local celestial meridian facing south from the 40°N latitude location.

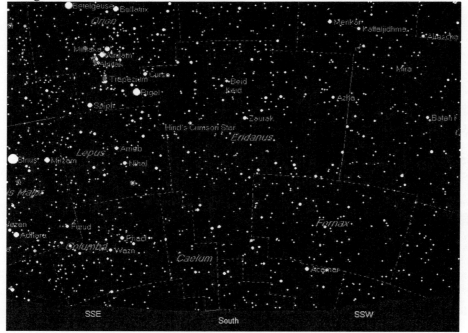

Figure 5-28 Local celestial meridian facing South at 40° N latitude

It is obvious, from looking at **Figure 5-28**, the local celestial meridian passes through the south point on the horizon. What is *not* so obvious, however, is the location of the celestial equator in this figure. Recall that the celestial equator is 90° from the north celestial pole; therefore looking south along the local celestial meridian, the celestial equator should be found at an altitude of 50° above the south point.

This can be easily illustrated for an observer located at 40° N latitude. First imagine you are standing outside and you draw a circle in the sky to represent the local celestial meridian. You must draw it from the north point, through the zenith, to the south point on the horizon. Next mark the location of the north celestial pole at 40° above the north point on the horizon. Then, measure an angle of 90° from the north celestial pole and mark its location on this circle. This point represents the location of the celestial equator.

Now draw a line from where you are standing on Earth to the north celestial pole mark and another line to the celestial equator mark. Now put in the angles. The angle left over is the one between the south point on the horizon and the celestial equator. This will be 50°! This is illustrated schematically for you in **Figure 5-29**.

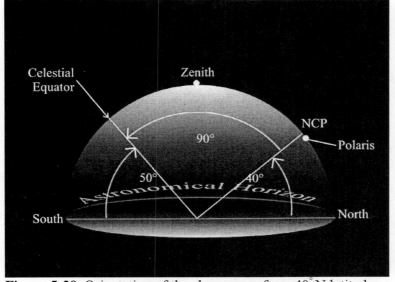

Figure 5-29 Orientation of the sky as seen from 40° N latitude

The large half-circle, in **Figure 5-29**, represents your local celestial meridian looking from the east.

In reality, the celestial equator is projected against the background stars along an east–west direction through the sky. It eventually passes through the east and west points on the horizon and disappears below your horizon. This can be seen in *TheSky* by clicking on either of the east or west buttons on the Orientation toolbar. There are two places on Earth where this is not true. They are the north and south geographic poles. At these two locations on Earth the celestial equator coincides with or lies on the astronomical horizon.

Figure 5-30, displays the equatorial grid. The blue circle labeled +00° in this figure represents the celestial equator in *TheSky*. Perpendicular to the celestial equator, and passing through the celestial poles, are vertical circles of right ascension spaced 1 hour (15°) apart.

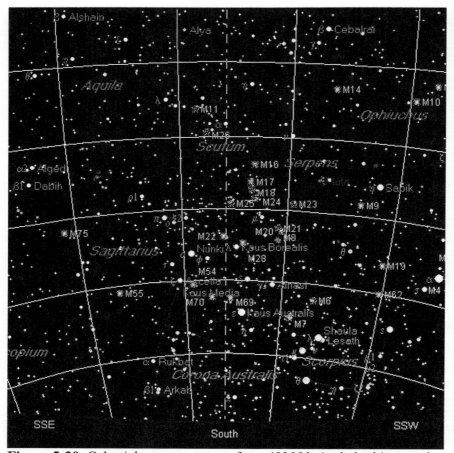

Figure 5-30 Celestial equator as seen from 40° N latitude looking south

Because the surface of Earth is spherical, it is interesting to note that the orientation of the sky changes the farther north or south you travel. Viewing the sky from different locations on Earth allows you to see how the orientation and positions of astronomical objects change.

Traveling north on Earth, the celestial equator appears lower in the sky while the north celestial pole appears higher. Traveling south on Earth, the celestial equator appears higher in the sky while the north celestial pole appears lower.

Knowing your location on Earth makes viewing the sky much easier and more fun. You can determine the sky's orientation for your location by estimating the altitude of the celestial pole and the celestial equator (Remember **Exercises A** and **B**?). If you are in the Northern Hemisphere, the altitude of the north celestial pole is the same as your latitude, whereas the altitude of the celestial equator is equal to 90° *minus* your latitude. The same is true for anyone living in the Southern Hemisphere, except it is the south celestial pole that is used there.

Many objects in the sky appear to rise and set throughout the day. Some, however, do not. These are the objects around the celestial poles. The motion of objects in the sky is also affected by your location on Earth. If you are located at the north or south geographic pole, objects in the sky appear to move parallel to your horizon. That is, nothing appears to rise or set at the geographic poles. Objects that never rise or set are known as *circumpolar objects*. However, if you are located at the geographic equator, then *every* object in the sky appears to rise and set.

Even though the sky appears to change its orientation as you travel to different locations on Earth, the coordinates of right ascension and declination in the sky remain the same. This is true no matter where you are on Earth. These coordinates are unaffected by the short-term motions of Earth. However, if you move ahead or backward in time hundreds or thousands of years, these coordinates are affected! This long-term effect on the equatorial coordinates is due to the Sun and Moon gravitationally tugging on Earth. The Earth responds to this tugging by precessing. Like a top, Earth's axis of rotation changes its orientation in space. One result of this motion is that Polaris will not always be our North Star. In 2500 B.C.E. the star α Draconis (Thuban) was the North Star. This star is the one that the Egyptians used as their North Star.

Finally, many people usually think of the vernal equinox, summer solstice, autumnal equinox, and the winter solstice as being points in time. That is, they represent the beginning of the seasons on Earth. These are associated with the approximate dates of March 21, June 21, September 21, and December 21, respectively. The seasons will be discusses in **Chapter 8**.

Astronomers, however, also think of these as points in space. At the time of the vernal equinox, the Sun appears to be crossing the celestial equator from the southern hemisphere into the northern hemisphere of the sky. On the summer solstice, the Sun appears to be at its greatest declination north of the celestial equator. At the time of the autumnal equinox, the Sun once again appears to be crossing the celestial equator, only this time it appears to be moving from the northern hemisphere into the southern hemisphere of the sky. On the winter solstice, the Sun appears to be at its greatest declination south of the celestial equator. And so it goes, year after year after year!

Thinking about the beginning of each season as a point in space allows astronomers to assign them equatorial coordinates. They are as follows:

Vernal equinox	R.A. = $0^H 0^M$	Declination = $0° 0'$
Summer solstice	R.A. = $6^H 0^M$	Declination = $+23° 26'$
Autumnal equinox	R.A. = $12^H 0^M$	Declination = $0° 0'$
Winter solstice	R.A. = $18^H 0^M$	Declination = $-23° 26'$

The equatorial coordinates of these points in space are illustrated in **Figure 5-31**.

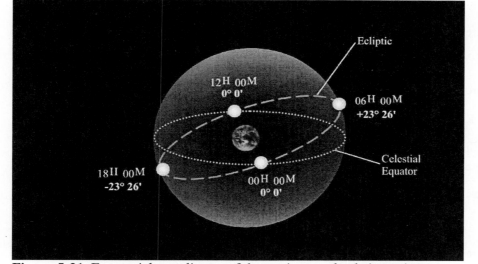

Figure 5-31 Equatorial coordinates of the equinox and solstice points

For telescopic observations the coordinates of right ascension and declination are readily used, and astronomical objects can be easily found. You need to determine the orientation of the equatorial system according to your location and time of your observation. *TheSky* is very proficient in keeping track of sky coordinates. It has an extensive database that lists the right ascension and declination of thousands of astronomical objects. It calculates their changes in altitude and azimuth caused by Earth's rotation. In addition, it updates the coordinates of right ascension and declination due to Earth's precession. This makes going outside and looking for things simple and more pleasant.

Chapter 5

TheSky Review Exercises

Before doing the following exercises, changes in *TheSky* are necessary.

TheSky Exercise 1: Observing *TheSky* at Ball State Observatory

Set the location, date, and time as follows:

Location: Ball State Observatory.
Date: October 25, 2006.
Time: 9:00 P.M. EST (Note: Ball State Observatory does not observe Daylight Savings Time).

1. What bright star lies near the north celestial pole? _____

2. If you could go ahead in time to the year 10,000 C.E., what bright star would be located close to the north celestial pole and would be our North Star? _____

3. If you could go back in time to the year 2,800 B.C.E., what bright star would be located close to the north celestial pole and would be our North Star? _____

** Set the date and time to that described above Question 1 for the remaining questions. **

4. What are the altitude and azimuth of Polaris?

    Altitude = ____° ____'   Azimuth = ____° ____'

5. What are the right ascension and declination of Polaris?

    R.A. = ____H ____M ____S   Declination = __ ___° ____' ____"

6. What constellation is located in the south on your local celestial meridian? _____

7. What are the altitude and azimuth of Alpheratz?

    Altitude = ____° ____'   Azimuth = ____° ____'

8. What are the right ascension and declination of Alpheratz?

    R.A. = ____H ____M ____S   Declination = __ ___° ____' ____"

9. In what constellation is Jupiter located? _____

10. What are the altitude and azimuth of Jupiter?

    Altitude = ____° ____'   Azimuth = ____° ____'

11. What are the right ascension and declination of Jupiter?

    R.A. = ___H ___M ___S   Declination = __ ___° ___' ___"

12. What are the right ascension and declination of M42?

$$\text{R.A.} = \underline{}^H \underline{}^M \underline{}^S \quad \text{Declination} = \underline{} \underline{}° \underline{}' \underline{}''$$

13. Is M42 currently visible? ____. If so, where would you have to look to see it?

_____

14. What are the altitude and azimuth of M42?

$$\text{Altitude} = \underline{}° \underline{}' \quad \text{Azimuth} = \underline{}° \underline{}'$$

15. Is M45 currently visible? _____

16. What type of object is M45? _____

17. What are the right ascension and declination of M45?

$$\text{R.A.} = \underline{}^H \underline{}^M \underline{}^S \quad \text{Declination} = \underline{} \underline{}° \underline{}' \underline{}''$$

18. What are the altitude and azimuth of M45?

$$\text{Altitude} = \underline{}° \underline{}' \quad \text{Azimuth} = \underline{}° \underline{}'$$

19. What astronomical object is located at R.A. = $00^H 42^M 45.2^S$ and Declination = +41° 6′ 19″?

_____

20. What type of object is the object in Question 19? _____

21. In what constellation is the bright star Sirius located? _____

22. How far away is Sirius? _____

23. What are the equatorial coordinates of IC 2118?

$$\text{R.A.} = \underline{}^H \underline{}^M \underline{}^S \quad \text{Declination} = \underline{} \underline{}° \underline{}' \underline{}''$$

24. What type of object is IC 2118? _____

25. In what constellation is the minor planet Vesta located? _____

26. How far is Vesta from the Sun? _____

27. How far is Vesta from Earth? _____

28. Find the Sun.

a. What are the right ascension and declination of the Sun?

$$\text{R.A.} = \underline{}^H \underline{}^M \underline{}^S \quad \text{Declination} = \underline{} \underline{}° \underline{}' \underline{}''$$

b. In what constellation is the Sun located on this date? _____

29. What is the classification of the object located at the coordinates of:

$$\text{R.A.} = 13^H 29^M 15^S \qquad \text{Declination} = +47° 11′ 54″? \underline{} \underline{}° \underline{}' \underline{}''$$

_____

30. What are the altitude and azimuth of the object in Question 29?

Altitude = ____° ____'   Azimuth = ____° ____'

TheSky Exercise 2: Observing *TheSky* in Atlanta, Georgia

Now make the following changes in *TheSky*. Set the location, date, and time as follows:

Location: Atlanta, Georgia
Date: October 25, 2006
Time: 9:00 P.M. EST

1. What bright star lies near the north celestial pole? _____

2. If you could go ahead in time to the year 10,000 C.E., what bright star would be located close to the north celestial pole and would be our North Star? _____

3. If you could go back in time to the year 2,800 B.C.E., what bright star would be located close to the north celestial pole and would be our North Star? _____

** Set the date and time to that described above Question 1 for the remaining questions. **

4. What are the altitude and azimuth of Polaris?

Altitude = ____° ____'   Azimuth = ____° ____'

5. What are the right ascension and declination of Polaris?

R.A. = ____H ____M ____S   Declination = __ ___° ____' ____"

6. What constellation is located to the South on your local celestial meridian?

_____

7. What are the altitude and azimuth of Alpheratz?

Altitude = ____° ____'   Azimuth = ____° ____'

8. What are the right ascension and declination of Alpheratz?

R.A. = ____H ____M ____S   Declination = __ ___° ____' ____"

9. In what constellation is Jupiter located? _____

10. What are the altitude and azimuth of Jupiter?

Altitude = ____° ____'   Azimuth = ____° ____'

11. What are the right ascension and declination of Jupiter?

R.A. = ___H ___M ___S Declination = __ ___° ___' ___"

12. What are the right ascension and declination of M42?

R.A. = ___H ___M ___S Declination = __ ___° ___' ___"

13. Is M42 currently visible? ____. If so, where would you have to look to see it?

_____

14. What are the altitude and azimuth of M42?

   Altitude = ____° ____'   Azimuth = ____° ____'

15. Is M45 currently visible? _____

16. What type of object is M45? _____

17. What are the right ascension and declination of M45?

   R.A. = ___H ___M ___S   Declination = __ ___° ___' ___"

18. What are the altitude and azimuth of M45?

   Altitude = ____° ____'   Azimuth = ____°____'

19. What astronomical object is located at R.A. = 00H 42M 45.2S and Declination = +41° 6' 19"?
   _____

20. What type of object is the object in Question 19? _____

21. In what constellation is the bright star Sirius located? _____

22. How far away is Sirius? _____

23. What are the equatorial coordinates of IC 2118?

   R.A. = ___H ___M ___S   Declination = __ ___° ___' ___"

24. What type of object is IC 2118? _____

25. In what constellation is the minor planet Vesta located? _____

26. How far is Vesta from the Sun? _____

27. How far is Vesta from Earth? _____

28. Find the Sun.

   a. What are the right ascension and declination of the Sun?

      R.A. = ___H ___M ___S   Declination = __ ___° ___' ___"

   b. In what constellation is the Sun located on this date? _____

29. What is the classification of the object located at the coordinates of:

   R.A. = 13H 29M 15S   Declination = +47° 11' 54"?

   _____

30. What are the altitude and azimuth of the object in Question 29?

   Altitude = ____° ____'   Azimuth = ____° ____'

31. What does the negative sign for the altitude mean? _____

32. Did you notice any differences in the answers in **Exercise 1** and **Exercise 2**? _____.

Were some the same? _____ Were some different? _____

Why were some of the answers different? _____

_____

Why were some of the answers the same? _____

_____

TheSky Exercise 3: Observing *TheSky* at the Tropic of Cancer

Now make the following changes in *TheSky*.
Set the location, date, and time as follows:

Location: Latitude: +23° 30′ 00″   Longitude: 85° 21′ 15″
Date: October 25, 2006
Time: 9:00 P.M. EST (+5)

1. What bright star lies near the north celestial pole? _____

2. If you could go ahead in time to the year 10,000 C.E., what bright star would be located close to the north celestial pole and would be our North Star? _____

3. If you could go back in time to the year 2,800 B.C.E., what bright star would be located close to the north celestial pole and would be our North Star? _____
    a. Now set the date and time to that described above Question 1 for the remaining questions.

4. What are the altitude and azimuth of Polaris?

     Altitude = ____° ____′   Azimuth = ____° ____′

5. What are the right ascension and declination of Polaris?

     R.A. = ____H ____M ____S   Declination = __ ___° ____′ ____″

6. What constellation is located to the South on your local celestial meridian?
    _____

7. What are the altitude and azimuth of Alpheratz?

     Altitude = ____° ____′   Azimuth = ____° ____′

8. What are the right ascension and declination of Alpheratz?

     R.A. = ____H ____M ____S   Declination = __ ___° ____′ ____″

9. In what constellation is Jupiter located? _____

10. What are the altitude and azimuth of Jupiter?

     Altitude = ____° ____′   Azimuth = ____° ____′

11. What are the right ascension and declination of Jupiter?

     R.A. = ___H ___M ___S   Declination = __ ___° ___′ ___″

12. What are the right ascension and declination of M42?

R.A. = ___^{H} ___^{M} ___^{S} Declination = __ ___° ___' ___"

13. Is M42 currently visible? ____. If so, where would you have to look to see it?
_____

14. What are the altitude and azimuth of M42?

Altitude = ____° ____'   Azimuth = ____° ____'

15. Is M45 currently visible? _____

16. What type of object is M45? _____

17. What are the right Ascension and declination of M45?

R.A. = ___^{H} ___^{M} ___^{S} Declination = __ ___° ___' ___"

18. What are the altitude and azimuth of M45?

Altitude = ____° ____'   Azimuth = ____° ____'

19. What astronomical object is located at R.A. = 00^H 42^M 45.2^S and Declination = +41° 6' 19"?
_____

20. What type of object is the object in Question 19? _____

21. In what constellation is the bright star Sirius located? _____

22. How far away is Sirius? _____

23. What are the equatorial coordinates of IC 2118?

R.A. = ___^{H} ___^{M} ___^{S} Declination = __ ___° ___' ___"

24. What type of object is IC 2118? _____

25. In what constellation is the minor planet Vesta located? _____

26. How far is Vesta from the Sun? _____

27. How far is Vesta from Earth? _____

28. Find the Sun.

a. What are the right ascension and declination of the Sun?

R.A. = ___^{H} ___^{M} ___^{S} Declination = __ ___° ___' ___"

b. In what constellation is the Sun located on this date? _____

29. What is the classification of the object located at the coordinates of:

R.A. = 13^H 29^M 15^S Declination = +47° 11' 54"?

_____

30. What are the altitude and azimuth of the object in Question 29?

Altitude = _____°_____′   Azimuth = _____°_____′

31. What does the negative sign for the altitude mean? _____

32. Did you notice any differences in the answers in Exercise 1 and Exercise 3? _____

Were some the same? _____ Were some different? _____

Why were some of the answers different? _____
_____

Why were some of the answers the same? _____
_____

33. Did you notice any differences in the answers in Exercise 2 and Exercise 3? _____

Were some the same? _____ Were some different? _____

Why were some of the answers different? _____
_____

Why were some of the answers the same? _____
_____

TheSky Exercise 4: Observing *TheSky* at 40° Latitude

Now make the following changes in *TheSky*.
Set the location, date, and time as follows:

Location: Latitude: 40° 11′ 15″   Longitude: 85° 21′ 15″
Date: December 25, 2007
Time: 5:00 P.M. EST (+5)

1. What are the altitude and azimuth of Polaris?

Altitude = _____°_____′   Azimuth = _____°_____′

2. What are the right ascension and declination of Polaris?

R.A. = _____ H _____ M _____ S   Declination = __ ___°_____′_____″

3. What constellation is located to the south on your local celestial meridian?
_____

4. What are the altitude and azimuth of Alpheratz?

Altitude = _____°_____′   Azimuth = _____°_____′

5. What are the right ascension and declination of Alpheratz?

R.A. = _____ H _____ M _____ S   Declination = __ ___°___′_____″

6. In what constellation is Jupiter located? _____

7. What are the altitude and azimuth of Jupiter?

                Altitude = ____° ____'   Azimuth = ____° ____'

8. What are the right ascension and declination of Jupiter?

                R.A. = ___^{H} ___^{M} ___^{S} Declination = __ ___° ___' ___"

9. In what constellation is Saturn located? _____

10. What are the altitude and azimuth of Saturn?

                Altitude = ____° ____'   Azimuth = ____° ____'

11. What are the right ascension and declination of Saturn?

                R.A. = ___^{H} ___^{M} ___^{S} Declination = __ ___° ___' ___"

12. What are the right ascension and declination of M42?

                R.A. = ___^{H} ___^{M} ___^{S} Declination = __ ___° ___' ___"

13. Is M42 currently visible? _____ If so, where would you have to look to see it?
_____

14. What are the altitude and azimuth of M42?

                Altitude = ____° ____'   Azimuth = ____° ____'

15. Is M45 currently visible? _____

16. What are the right ascension and declination of M45?

                R.A. = ___^{H} ___^{M} ___^{S} Declination = __ ___° ___' ___"

17. What are the altitude and azimuth of M45?

                Altitude = ____° ____'   Azimuth = ____° ____'

18. What is located at R.A. = $00^H 36^M 45.2^S$ Declination = +42° 16' 19"?
_____

19. What type of object is the object in Question 18? _____

20. In what constellation is the bright star Procyon located? _____

21. How far away is Procyon? _____

22. Find the Sun. Center it in the desktop window.

   a. What are the Sun's right ascension and declination?

                R.A. = ___^{H} ___^{M} ___^{S} Declination = __ ___° ___' ___"

   b. What are the Sun's Altitude and Azimuth?

                Altitude = ____° ____'   Azimuth = ____° ____'

c. What time does it appear to transit the local celestial meridian today? _____:_____

23. Find the Moon. Center it in the desktop window.

    a. What are the Moon's right ascension and declination?

        R.A. = ___^H ___^M ___^S   Declination = __ ___° ___′ ___″

    b. What are the Moon's altitude and azimuth?

        Altitude = ____° ____′   Azimuth = ____° ____′

    c. At what time does it appear to transit the local celestial meridian today? _____:_____

24. Do you suppose there was a solar eclipse today? _____

25. If so, what time does it occur? _____:_____

TheSky Exercise 5: Observing *TheSky* at the Geographic Equator

Now make the following changes in *TheSky*.
Set the location, date, and time as follows:

Location: Latitude: 0° 0′ 0″   Longitude: 85° 21′ 15″
Date: December 25, 2007
Time: 5:00 P.M. EST (+5)

1. What are the altitude and azimuth of Polaris?

        Altitude = ____° ____′   Azimuth = ____° ____′

2. What are the right ascension and declination of Polaris?

        R.A. = ____^H ____^M ____^S   Declination = __ ___° ____′ ____″

3. What constellation is located to the South on your local celestial meridian?
    _____

4. What are the altitude and azimuth of Alpheratz?

        Altitude = ____° ____′   Azimuth = ____° ____′

5. What are the right ascension and declination of Alpheratz?

        R.A. = ____^H ____^M ___^S   Declination = __ ___° ____′ ____″

6. In what constellation is Jupiter located? _____

7. What are the altitude and azimuth of Jupiter?

        Altitude = ____° ____′   Azimuth = ____° ____′

8. What are the right ascension and declination of Jupiter?

        R.A. = ___^H ___^M ___^S   Declination = __ ___° ___′ ___″

9. In what constellation is Saturn located? _____

10. What are the altitude and azimuth of Saturn?

   Altitude = ____° ____'   Azimuth = ____° ____'

11. What are the right ascension and declination of Saturn?

   R.A. = ___^{H} ___^{M} ___^{S} Declination = __ ___° ___' ___"

12. What are the right ascension and declination of M42?

   R.A. = ___^{H} ___^{M} ___^{S} Declination = __ ___° ___' ___"

13. Is M42 currently visible? _____ If so, where would you have to look to see it?
   _____

14. What are the altitude and azimuth of M42?

   Altitude = ____° ____'   Azimuth = ____° ____'

15. Is M45 currently visible? _____

16. What are the right ascension and declination of M45?

   R.A. = ___^{H} ___^{M} ___^{S} Declination = __ ___° ___' ___"

17. What are the altitude and azimuth of M45?

   Altitude = ____° ____'   Azimuth = ____° ____'

18. What astronomical object is located at R.A. = 22^H 30^M 02^S and Declination = –20°40' 39"?
   _____

19. What type of object is the object in Question 18? _____

20. In what constellation is the bright star Procyon located? _____

21. How far away is Procyon? _____

22. Find the Sun. Center it in the desktop window.

   a. What are the Sun's right ascension and declination?

      R.A. = ___^{H} ___^{M} ___^{S} Declination = __ ___° ___' ___"

   b. What are the Sun's altitude and azimuth?

      Altitude = ____° ____'   Azimuth = ____° ____'

   c. What time does it appear to transit the local celestial meridian today? _____:_____

23. Find the Moon. Center it in the desktop window.

   a. What are the Moon's right ascension and declination?

      R.A. = ___^{H} ___^{M} ___^{S} Declination = _ ___° ___' ___"

b. What are the Moon's altitude and azimuth?

Altitude = _____° _____'   Azimuth = _____° _____'

c. What time does it appear to transit the local celestial meridian today? _____:_____

24. Do you suppose there was a solar eclipse today? _____

25. If so, what time did it occur? _____:_____

Chapter 5

TheSky Review Questions

Part A

Answer these basic questions about Earth and the geographic coordinates:

1. The two points on Earth intersected by its axis of rotation are known as the _____ and _____.

2. What is the name of the line drawn on Earth dividing it into two equal halves, north and south? _____.

3. Great circles are circles that pass through the North and South Poles on Earth. The Earth is divided into 24 equal parts east to west along its equator. The angle between any two of these circles is ____ hour or ____°.

4. Any two points on the surface of Earth that are located on *one* of these great circles and is located in the *same* hemisphere will have the same _____.

5. All points on the surface of Earth that have latitude circles that are not zero are known as _____ circles.

6. What is the latitude of the north geographic pole? _____°

7. What is the longitude of the north geographic pole? _____°

Part B

Answer these basic questions about the sky coordinates

1. Which coordinate in the equatorial coordinate system is like longitude in the geographic coordinate system? _____

2. Which coordinate in the equatorial coordinate system is like latitude in the geographic coordinate system? _____

3. Which coordinate in the horizon coordinate system is similar to longitude in the geographic coordinate system? _____

4. Which coordinate in the horizon coordinate system is similar to latitude in the geographic coordinate system? _____

5. What units are used to measure the coordinates of right ascension (R.A.)? _____

6. What direction in the sky is the coordinate of R.A. measured? _____

7. What is the range in values of right ascension? _____

8. What is the name of the point in the sky that defines the secondary circle in the equatorial coordinate system? _____

Chapter 6

Motions in *TheSky*

Observing the stars continues to be an essential activity for many cultures today. For millennia seafarers have used the stars to navigate by, something that continues even into modern times.

Being familiar with how stars are designated and how to find their location in the sky is important in observational astronomy. Now it is time to focus attention on the motions observed in the sky. In ancient times, most astronomers believed that such motions were actually caused by the objects moving around us. This error led some of them to formulate intricate cosmologies to explain these motions. Today, we understand the motions and can explain them correctly.

There are basically two kinds of motions observed in the sky, *short-term* and *long-term* motion. The short-term motion is called *daily* or *diurnal* motion and is responsible for the daily rising and setting of the Sun, Moon, planets, stars, and everything else in the sky. A consequence of short-term motion is that most celestial objects generally move in a westerly direction across the sky during the course of a day. That is, they rise in the east and set in the west. Short-term motion has a period of time that is on the order of a day or less (hours, minutes, seconds).

As mentioned in the last chapter, the entire celestial sphere rotates about two fixed points in the sky, the north and south celestial poles. This motion, however, is an apparent motion. Daily motion is actually a result of Earth rotating on its axis. Diurnal motion is responsible for the time-dependent change of the horizon coordinates and the ever-changing positions of objects in the sky. The rate at which objects move in the sky is 15° per hour from east to west.

In this chapter, we use many multimedia files that may be thought of as sky windows. These files should have already been downloaded into the folder C:\Program Files\Software Bisque\ TheSky\Users\Documents\Multimedia. This folder was created during the installation procedure described in Chapter 1. The URL for the multimedia sky files is: http://www.brookscole.com/product/0534390722. These files are labeled with a "filename" and an extension labeled "sky." These sky window files have been created to help you visualize and demonstrate the motions you see in the sky.

The following exercises are designed to help you open the multimedia files. Then you can view both short-term and long-term motions in the sky. The first few exercises give explicit details in how to open and view the sky windows. Later exercises indicate which files to open and view.

The Sun's apparent rising on the eastern horizon is perhaps the most obvious example of diurnal motion. The multimedia file named SunRise.sky is a sky window that illustrates this motion. Exercise A illustrates how to open the multimedia file SunRise.sky and view the apparent sunrise in *TheSky*.

Exercise A: Observing the Diurnal Motion of the Sun

1. Start *TheSky*.

2. Click the *Open* file button () on the Standard toolbar, or, go to File, then Open.

3. Double-click on User, then Documents, then Multimedia. An Open window appears like the one displayed in Figure 6-1. All the multimedia files are displayed in this folder.

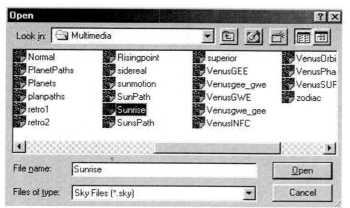

Figure 6-1 Opening the multimedia file SunRise.sky

4. Select the proper demonstration filename—in this case SunRise.sky. The extension is listed on the "Files of type" line at the bottom of the window (*.sky).

5. Click Open to open the file.

6. After opening the file, click the [▶▶] button and simply watch. If the motion is too fast, then click the [▲] button and then use the [▶] button to view the sky window at a slower rate. Notice that the direction in which the observer is facing is east.

7. To reset the window to the beginning and view this motion again, click the [▲] button.

8. To close this file, click the Open file button ([📂]) again and choose the file Normal.sky, or continue with Exercise B. It is not necessary to save this multimedia file.

Exercise B: Observing the Diurnal Motion of the Stars, Part I

1. Click the Open File button ([📂]) on the Standard toolbar, or, go to File, then Open at the top of the sky window.

2. An Open window reappears like the one shown in Figure 6-1 with the multimedia files displayed. Now double-click on the multimedia file named StarsMotionS.sky.

3. After opening the file, click the [▶▶] button and watch.

4. Notice the *view* is looking *south*. The stars appear to move from east to west. If you look closely, the Moon and a few planets can also be seen in this window.

5. If the motion is too fast, click the [▲] button and use the [▶] button to view this sky window at a slower rate.

6. To close this file, click the Open file button ([📂]) again and choose the file Normal.sky, or continue with Exercise C. It is not necessary to save this multimedia file.

Exercise C: Observing the Diurnal Motions of the Stars, Part II

1.   Click ⬀ on the Standard toolbar, or, go to File, then Open at the top of the sky window.

2.   Double-click on the multimedia file named StarsMotionE.sky.

3.   After opening the file, click the ⏭ button and watch.

4.   This file is the same as the one in Exercise B except the view is looking east. Notice in this sky window that all the stars appear to be rising at an angle to the eastern horizon. As before, as the sky continues to darken more stars become visible.

5.   If the motion is too fast, then click the ⏏ button and use the ▶ button to view this sky window at a slower rate.

6.   To close this file, click the Open file button (⬀) again and choose the file Normal.sky, or continue with Exercise D. It is not necessary to save this multimedia file.

Exercise D: Observing the Diurnal Motions of the Stars, Part III

1.   Click ⬀ on the Standard toolbar, or, go to File, then Open at the top of the sky window.

2.   Double-click on the multimedia file named StarsMotionW.sky.

3.   After opening the file, click the ⏭ button and watch.

4.   This file is the same as the previous windows except the view is looking west. Notice the stars appear to be setting below the western horizon at an angle. As the sky continues to darken more stars become visible. If you watch this window long enough, the Moon and later a few planets also appear to set.

5.   If the motion is too fast, then click the ⏏ button and use the ▶ button to view this sky window at a slower rate.

6.   To close this file, click the Open file button (⬀) again and choose the file Normal.sky or continue with Exercise E. It is not necessary to save this multimedia file.

Exercise E: Observing the Diurnal Motions of the Stars, Part IV

1.   Click ⬀ on the Standard toolbar, or, go to File, then Open at the top of the sky window.

2.   Double-click on the multimedia file named StarsMotionN.sky.

3.   After opening the file, click the ⏭ button and watch.

4.   This file is the same as the previous windows except that the view is looking north.

In the middle of the sky window is the bright star Polaris, the North Star. Notice that stars in the sky window appear to move counterclockwise about Polaris, or more correctly about the north celestial pole. Notice, too, that the circumpolar stars never go below the horizon.

As the sky continues to darken, more and more stars become visible. Their motions become very evident after you observe for some time. Stars *above* Polaris appear to move from east to west (to the left). Stars *below* Polaris appear to move from west to east (to the right). Stars to the *left* of Polaris appear to move *downward*. Stars to the *right* of Polaris appear to move *upward*.

The overall motion about the north celestial pole is in a counterclockwise direction. The counterclockwise motion you observe in this sky window is caused by the Earth itself rotating counterclockwise (west to east) on its axis. The time skip has been set to 5-minute intervals.

5. If the motion is too fast, then click the [⏏] button and use the [▶] button to view this sky window at a slower rate.

6. To close this file, click the Open file button ([📂]) again and choose the file Normal.sky, or close it. It is not necessary to save this multimedia file.

Exercises B, C, D, and E could all have been done as one exercise. The more often you use this software, the easier this sequence will become.

In ancient times people noted different types of objects in the sky. They called the constellations simply the "fixed" stars. These stars didn't move in the sky except from east to west. Their patterns remained the same year after year for centuries. Ancient astronomers took these things very seriously, because the stability probably had a calming effect on their thinking and their culture. There was no chaos in this universe, only order.

Long-term motion is often referred to as *annual* motion. It corresponds to time periods longer than 24 hours (weeks, months, and even years). Observing the nighttime sky over longer periods of time, you will notice, as the ancients did, that not all stars remain fixed. Some of them move relative to the "fixed" stars or constellations. The ancient Greeks called these moving objects the *planetes* or "wandering stars." *Planetes* is the origin of the word planets. There are five naked-eye planets, which move along well-defined paths near the ecliptic. The multimedia file named Planets.sky in Exercise F illustrates how these objects appear to move in Earth's sky.

Exercise F: Observing the Planets' Long-Term Motion

1. Start *TheSky,* or if it is already running click the [📂] button on the Standard toolbar.

2. Now select the multimedia file named Planets.sky.

3. Click Open to open the file.

4. After opening the file, click the [⏭] button. Again, if the motion is too fast, then click the [⏏] button and use the [▶] button.

5. To reset the window to the beginning, at any time, click the [⏏] button once again.

6. To close this file, click the Open file button ([📂]) and choose the file Normal.sky or continue with Exercise G. It is not necessary to save this multimedia file.

The most obvious object in the night sky is the Moon. After a few nights of observing, you soon realize that it too moves relative to the constellations. In ancient times, the Moon was also considered a planet. Ancient Greek and Renaissance astronomers alike took the definition of *planetes* literally when it came to the objects in the sky: the planets "wandered." The multimedia file named MoonMotion.sky illustrates the Moon's motion in the sky over a period of several days. Observe the motion in the next exercise carefully! Pay particular attention to the path that the Moon makes relative to the stars and compared it to that of the Sun's (teal line). Exercise G demonstrates the Moon's long-term motion in the sky.

Exercise G: Observing the Moon's Long-Term Motion

1. Start *TheSky,* or if it is already running click the ⬚ button on the Standard toolbar.

2. Now select the multimedia file named MoonMotion.sky.

3. Click <u>O</u>pen to open the file.

4. Click the ⬚ button to watch the motion in this file. Click the ⬚ button when finished.

5. Close the file. Once again, it is not necessary to save this multimedia file.

If you observe the Sun over a period of several months, you notice that it too moves relative to the "fixed" stars. Its motion is along a well-defined path known as the ecliptic. During the course of a year, the Sun appears to move through thirteen constellations in the sky. The multimedia file named SunMotion.sky demonstrates this eastward motion along the ecliptic. In *TheSky*, a teal-colored line in all sky windows represents the ecliptic. This is the default color setting and may be changed at any time (see Appendix E).

The Sun's motion is both north and south of the celestial equator. Exercise H illustrates the Sun's motion in the sky. Remember that this motion is an apparent motion. The Sun's apparent motion along the ecliptic is caused by Earth revolving around the Sun, not the Sun revolving around Earth as most ancients believed.

Exercise H: Observing the Sun's Long-Term Motion

1. Start *TheSky,* or if it is already running click the ⬚ button on the Standard toolbar.

2. Now select the multimedia file named SunMotion.sky.

3. Click <u>O</u>pen to open the file.

4. Click the ⬚ button to watch this file. Click the ⬚ button when finished.

5. Close the file. It is not necessary to save this multimedia file.

The constellations that are centered on the ecliptic are known as the *zodiacal constellations* or the *zodiac*. Exercise I illustrates the motion of the Sun through these constellations.

Exercise I: Observing the Sun's Motion through the Zodiac

The constellation names in this exercise are very familiar (Sagittarius, Capricornus, Aquarius, Pisces, Aries, Taurus, and so on).

1. Start *TheSky,* or if it is already running click the ⬜ button on the Standard toolbar.

2. Now select the multimedia file named Zodiac.sky.

3. Click Open to open the file.

4. Click the ⏭ button to watch this file. Click the ⬛ button when finished.

5. Using the directional button (⬅) on the Orientation toolbar, follow the apparent path of the Sun around the sky. Make sure you move through a full circle.

6. Close the file when finished. It is not necessary to save this multimedia file.

The planets and the Moon are always found somewhere along and within ±8° of the Sun's apparent path. The file named PlanetPaths.sky illustrates the motion of all these objects near the ecliptic. This is quite simple to understand when you think of our Solar System as being essentially a "flat" system. Exercise J illustrates the motion of Solar System objects in Earth's sky.

Exercise J: Observing the Motion of Sun, Moon, and Planets in the Sky

1. Start *TheSky,* or if it is already running click the ⬜ button on the Standard toolbar.

2. Now select the multimedia file named PlanetPaths.sky.

3. Click Open to open the file.

4. Click the ⏭ button to watch this file. Click the ⬛ button when finished.

5. It is not necessary to save this multimedia file.

Another short exercise that can be done while in this sky window, is to use *TheSky's* 3D Solar System Mode option. This mode allows you to view the Solar System in a three-dimensional (3D) representation at any time. It depicts all the planets in their respective orbits and their location with respect to the Sun and Earth at the time of your observation. Exercise K shows how to view the Solar System in a 3D mode.

Exercise K: Observing a 3D Model of Our Solar System

1. Using the sky window in Exercise J, click on the "3D Solar System Mode" button (⬜).

2. Now click the ⏭ button to put the planets into motion.

3. Notice that the motion of all the objects in this sky window is counterclockwise about the Sun. It is helpful here to use the ![zoom in button] button on the Orientation toolbar to watch the motion.

4. Click the button on the right-hand side of the window and scroll down with the mouse until you observe our Solar System from the plane of the Earth's orbit.

5. Notice that with the exception of Pluto, all the planetary orbits are very close to Earth's orbit.

6. Click the ![zoom button] button to observe planets close to the Sun.

7. Click on the "3D Solar System Mode" button (![3D button]) again to return to the sky mode.

8. Close the file. It is not necessary to save this multimedia file.

All the counterclockwise motion observed in the previous exercise translates to eastward motion as viewed from Earth's surface. In other words, long-term motion of the Sun, Moon, and planets is generally in an eastward direction as projected against the background stars in Earth's sky. The *eastward motion* of planets is often referred to as *direct motion*.

The Moon takes about a *month* to complete an entire circuit around the celestial sphere, whereas it takes the Sun about a *year* to accomplish the same thing. Planets, in contrast, moving at different rates through the sky, have time periods that vary in duration. The duration to complete a circuit around the celestial sphere with respect to the stars depends upon their distance from the Sun. No matter which moving object is observed, each has its own time period with respect to the background stars. This period of time is known as the planet's *sidereal period*.

Sometimes planets are observed moving westward in the sky during this long-term cycle. This *westward* motion is called *retrograde motion*. Several multimedia files have been rendered to display both direct and retrograde motions of the planets.

Figure 6-2 illustrates the retrograde motion of the planet Jupiter as projected against the background stars. In this figure, the trail was added to reproduce the motion that is observed over a period of several months.

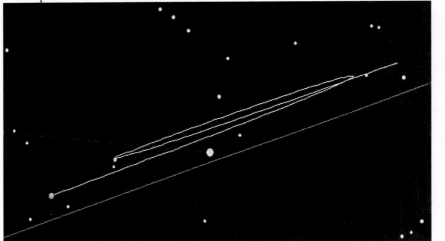

Figure 6-2 Jupiter retrograding

The multimedia file named JupiterRetro.sky, for example, illustrates Jupiter's retrograde motion in Libra from approximately November 27, 2005, to about October 29, 2006. Exercises L and M, which follow, will demonstrate motions of several planets as observed in Earth's sky projected against the background stars. Watch these motions carefully!

Exercise L displays planets moving eastward or in direct motion. The planets Mercury and Venus, however, do appear to exhibit retrograde motion in this sky window. Exercise M demonstrates the appearance of Jupiter's motion in Earth's sky when Jupiter is retrograding. By clicking on the "3D Solar System Mode" button, you can view the positions of Earth and Jupiter in the Solar System during this period of time.

Exercise L: Observing Retrograde Motion, Part I

1. Open the multimedia file named Retro1.sky then click the ▶▶ button.

2. Periodically clicking the ■ button displays the planetary labels.

3. Close the file or continue with Exercise M. It is not necessary to save this multimedia file.

Exercise M: Observing Retrograde Motion, Part II

1. Open the multimedia file named Retro2.sky then click the ▶▶ button.

2. Periodically clicking the ■ button displays the planetary labels.

3. Click the "3D Solar System Mode" button ( ) to observe Jupiter's position relative to the Earth.

4. Close the file. It is not necessary to save either of these multimedia files.

Of all these motions, the Sun's motion along the ecliptic and the retrograde motion of the planets are both apparent motions. The Sun's eastward motion along the ecliptic is actually caused by Earth revolving around it. The north–south motion of the Sun along this path is caused by the 23.5° tilt of Earth's axis. This motion can be viewed by reviewing the sky window SunMotion.sky.

The tilt of Earth is what makes the Sun appear to move north and south of the celestial equator. This, in turn, is responsible for the change of the apparent rising point of the Sun on the eastern horizon throughout the course of a year. It is also the main cause of the seasons on Earth (see Chapter 9).

The Sun's apparent rising point moves both northward and southward along the eastern horizon. *TheSky* can easily demonstrate this motion. The multimedia file named RisingPoint.sky was designed to do this. When you face east, you must first locate where the ecliptic intersects the eastern horizon. This is easy, just look for the zodiacal constellations!

If you imagine now that the Sun is located at this point of intersection, as it appears to rise, then this point *is* the apparent rising point of the Sun on the eastern horizon. After watching over a period of a few months, you soon realize that this intersection point moves both northern and southward along the eastern horizon. Ancient astronomers observed this phenomenon thousands of years ago and constructed accurate calendars based on this very observation. Stonehenge, in England, is a stone calendar that marks the location of the Sun's rising point on the eastern horizon at specific times of the year.

Exercise N has been designed to illustrate this motion. When the Sun is located at the northernmost or southernmost point on the horizon, the Sun appears to rise on the summer solstice or winter solstice, respectively. The word *solstice* is a Greek word meaning "*sun standing still.*" That is, the Sun appears to stop its northerly and southerly motion along the eastern horizon at these two points. When the Sun's apparent rising point is due east, the Sun appears to rise due east on the vernal or autumnal equinox, respectively.

Exercise N: Observing the Sun's Rising Point on the Eastern Horizon

1. Open the multimedia file named RisingPoint.sky.

2. Click the East button (⊞), if you are not already facing that direction.

3. Find where the ecliptic intersects the eastern horizon. Click the ⏩ button and watch this intersection point (teal-colored line) move northward, then southward along the horizon.

4. Close the file. It is necessary to save this multimedia file.

Because the retrograde motion of planets is an apparent motion, it may seem a bit more difficult to determine its cause. The combined motion of Earth and the planets causes this motion. A similar thing is often observed while passing a slower moving vehicle on a four-lane highway. As you look out the window at the car being passed, the other vehicle appears to be moving backward with respect to distant trees, houses, or billboards. The same is true for planets. Faster moving planets overtake and pass slower moving ones. When they do, the slower moving planet appears to back up, or move westward, in Earth's sky as seen projected against the background stars, as in Figure 6-2.

The multimedia file named EarthMars1.sky shows the orbits of Earth and Mars in a 3D mode as they revolve around the Sun. There are two things to notice when viewing this sky window. The first is that both planets move in the same direction around the Sun. The second is that Earth is moving faster than Mars and overtakes it. In fact, this event occurs quite regularly in our Solar System. Earth overtakes Mars approximately every 779 days.

This period of time is referred to as a Mars' synodic period. That is, every 2.1 years Earth catches up with and overtakes Mars in its orbit. At the time this event occurs, Mars is 180° from the Sun or in a configuration that is known as *opposition*. During this time it appears to *retrograde* or *move westward* in Earth's sky. Exercise O illustrates the position and motion of both Earth and Mars in a 3D representation of the Solar System.

Exercise O: Observing the Motion of Earth and Mars about the Sun

1. Open the multimedia file named EarthMars1.sky.

2. After opening the file, click ⏩ and watch the motion just described.

3. Close the file. It is not necessary to save this multimedia file.

The multimedia file named EarthMars2.sky displays how the motion of Mars appears in Earth's sky projected against the background stars. The time interval is approximately the same as it was in the EarthMars1.sky multimedia file. Exercise P illustrates the motion of Mars in Earth's sky during the time when Mars is in opposition. The time interval, during which the retrograde motion occurs, is from approximately May 1, 2001, to July 27, 2001. The next time that Mars appears to retrograde will be sometime during the middle of July 2003.

Exercise P: Observing Retrograde Motion of Mars in Earth's Sky

1. Open the multimedia file named EarthMars2.sky.

2. After opening the file, click ⏩ and watch the motion as previously described.

3. Close the file. It is not necessary to save this multimedia file.

In fact, all the outer planets appear to retrograde in this fashion but their synodic periods are different. Jupiter's retrograde motion is illustrated in the multimedia file named JupiterRetro.sky. Viewing this multimedia file will be left as an exercise for the student.

Exercise Q helps you set the sky window to view the retrograde motion of Mars in the year 2003.

Exercise Q: Observing Mars Retrograde in July 2003

1. Re-open the multimedia file used in Exercise P.

2. Click Data on the toolbar located at the top of the sky window.

3. Click Site Information. The location setting can be of your own choosing.

4. Now click the Date and Time tab.

5. Set Date: July 15, 2003.
    Time: 2:30 A.M. EST
    Daylight saving adjustment option: "North America."

6. Click Apply, then Close.

7. Find Mars. To do this, click Edit on the toolbar located at the top of the sky window, then click Find. You may also right-click the mouse anywhere in the sky window then click Find. After the Object Information window opens, center Mars.

8. Click the ▶▶ button and watch.

9. Does Mars appear to retrograde? _____. On what date does Mars appear to begin to retrograde? _____. If you change your location on Earth, how will this affect what will be observed? _____.

10. Try it! Set your location to Moscow, CIS. Click Apply, then Close.

11. Click the ▶▶ button, and observe. Are there any changes? _____.

12. You can save this file by clicking the Save button (🖫). First, a word of caution!

Caution: Before saving this file you should change the filename!

By clicking the Save file (🖫) button, the multimedia sky window that you just viewed will replace the multimedia file named EarthMars2.sky. To avoid this problem, you must use the Save As option instead. At the top of the sky window, click File on the toolbar and then Save As. Type in the dialog box provided "RetroMars72003.sky" to rename this file. This is just an example of a filename; you may choose any filename as long as it is not a filename that already exists in the Multimedia folder.

13. Now close the file.

Today we recognize the Sun as a star, the Moon as a satellite of Earth, and the planets as individual worlds orbiting the Sun. They are members of our Solar System.

To observe them takes some time and patience but distinguishing planets in the sky is really not as difficult as you might think. Usually they appear as "bright" stars in the sky. After you observe for a few weeks, you may have seen that the object moves relative to the distant stars if it's a planet. This is the most tried and true method.

There is another easy way to ascertain whether an object is a planet or a bright star. First, find its location in the sky with respect to the ecliptic. A "bright" star along the ecliptic may be a planet. Once again, if you observe it for a few weeks and it moves, it definitely is a planet!

The most reliable resource for locating planets is *TheSky*. It calculates the positions of all Solar System objects with amazing accuracy. You can find planets easily for any year or any time of the year using this software. Whether or not the planets are visible remains to be "seen!"

Classifying Planets:

To make sense of planetary motions, it can help to have an understanding of the way planets are classified, how they move, and where they are located in our Solar System. There are several ways in which we classify planets their distance from the Sun is just one way.

Using Earth's distance as a "standard" distance in the Solar System, you find that there are two types of planets. The first type is called an *inferior planet*. An inferior planet is a planet whose orbit lies *inside* Earth's orbit. That is, its semimajor axis or average distance from the Sun is *less than* 1 astronomical unit (1 AU), or about 93,000,000 miles.

There are two inferior planets, Mercury and Venus. The multimedia file that is named Inferior.sky shows the orbits and motions of Mercury and Venus with respect to Earth and the Sun. Inferior planets all have periods of revolution of less than one year. Exercise R shows the motion of Mercury, Venus, and Earth, as viewed from a location outside and above the Solar System. After you open this multimedia file, press the Go Forward button to watch the motion of these planets. Using the scroll button at the right-hand side of the window will let you change your viewing perspective from above to below the plane of the Solar System.

Exercise R: Observing the Orbits and Motions of Mercury and Venus

1. Open the multimedia file named Inferior.sky.

2. After you open the file, click ⏩ and observe the motion.

3. Be sure and use the scroll bar at the right-hand side of the sky window to change your viewing perspective.

4. Close the file. It is not necessary to save this multimedia file.

The second type of planet is called a *superior planet*. Superior planets are planets whose orbits *lie outside* Earth's orbit. Their semimajor axes are *greater than* 1 AU and their periods of revolution are greater than one year. There are six superior planets, Mars, Jupiter, Saturn, Uranus, Neptune, and Pluto. Of these, Mars, Jupiter, and Saturn are bright enough to be seen with the naked eye. The multimedia file named Superior.sky shows the orbits and motions of the superior planets with respect to Earth and the Sun as viewed from outside and above the Solar System. Exercise S shows the motion of only the planets beyond Earth's orbit. After you open this file, press the Go Forward button to watch the motion of these planets. Using the scroll button at the right-hand side of the window will once again let you change your viewing perspective from above to below the plane of the Solar System.

Exercise S: Observing the Orbits and Motions of the Superior Planets

1. Open the multimedia file named Superior.sky.

2. After opening the file, click ▶▶ and observe the motion previously described.

3. Move the scroll bar at the right edge of the sky window down. Eventually you pass into the plane of the Solar System. Look at the orbits! How do they appear? _____
    Can you draw any conclusions about the "shape" of our Solar System? _____

4. Close the file. It is not necessary to save this multimedia file.

All the planets move around the Sun in a counterclockwise direction when viewed from above the Solar System. As stated earlier, this motion around the Sun translates to eastward motion in Earth's sky. The sidereal period is a measurement of the time it takes a planet to make one complete revolution around the Sun with respect to the stars. The sidereal periods for the planets are in the same order as their distance from the Sun. The multimedia file named Sidereal.sky displays all the naked-eye planets moving around the Sun. Exercise T illustrates this motion.

Exercise T: Observing the Motions of the Naked-Eye Superior Planets

1. Open the multimedia file named Sidereal.sky.

2. Click ▶▶ and observe the motion.

3. Use the "Zoom Out" button (🔍) on the Orientation toolbar, to view planets beyond Jupiter.

4. Move the scroll bar at the right edge of the sky window down again into the plane of the Solar System. Look at the orbits! Are they about in the same plane? _____. Would you Say that our Solar System is a flat system? _____

5. Close the file. It is not necessary to save this multimedia file.

A German mathematician figured all this motion out hundreds of years ago. In 1609 Johannes Kepler published the book entitled *Astronomia Nova* (*The New Astronomy*). In this book he described planetary motion in the Solar System and formally presented his first two laws of planetary motion.

His first law, known as the *law of ellipses*, described the shape of planetary orbits. After working many years with positional data for the planets, he found that they move in elliptical orbits around the Sun. He also realized that the Sun's position is not exactly in the middle of each planet's orbit. It is off center at a point he called the *focus point*. A schematic diagram of Kepler's first law is shown in Figure 6-3.

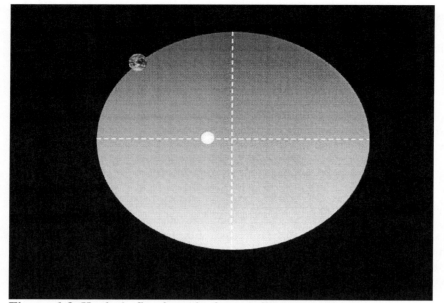

Figure 6-3 Kepler's first law, the *law of ellipses*

Kepler's second law of planetary motion describes the motions of the planets themselves. It is known as the *law of areas*. He found that if you connect a line between the Sun and a planet, and let this line move as the planet moves around the Sun, it sweeps out *equal areas* in space in *equal time intervals*. The line, connecting the Sun and a planet, is called a *radius vector*.

Figure 6-4 illustrates a radius vector connecting the Sun and a planet at points A, B, C, and D. Imagine, as Kepler did, the planet moving in different segments (A to B or C to D, and so on) in its orbit around the Sun. In some time interval, say t, the planet moves from point A to point B in its orbit, sweeping out area P in space. At some later time, same t, the planet moves from point C to point D in its orbit, sweeping out area A. If you measure these two areas carefully you find, as Kepler did, that they are equal!

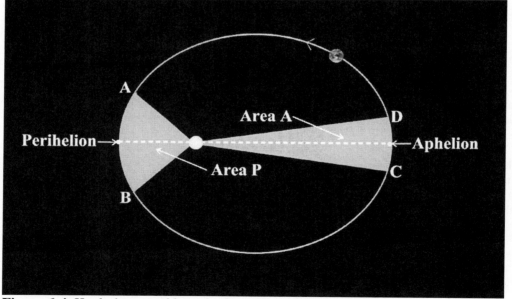

Figure 6-4 Kepler's second law, the *law of areas*

For the second law to be valid, Kepler realized, a planet speeds up and slows down as it orbits the Sun. It moves faster from point A to point B than from point C to point D. A planet travels fastest in its orbit when it is halfway between points A and B and slowest in its orbit when it is halfway between points C and D.

Kepler defined the two points in a planet's orbit where it is *closest* and *farthest* from the Sun as the *perihelion* and *aphelion*, respectively. So as planets orbit the Sun, their distance varies and they speed up and slow down. The closest distance for the planet lies exactly halfway between points A and B and the farthest distance lies exactly halfway between points C and D in Figure 6-4.

If a horizontal line is drawn from the closest distance to the farthest distance in Figure 6-4, this represents the largest diameter of the ellipse and is defined as the *major axis* of the ellipse. A vertical line drawn through the center of the ellipse, dividing the major axis into two equal parts represents the smallest diameter of the ellipse. This line is defined as the *minor axis* of the ellipse. These two lines are axes of symmetry in the ellipse.

Kepler determined that the average distance that a planet is from the Sun is equal to the length of its *semimajor axis* (half of the largest diameter). He also defined a new unit of distance for the Solar System. This distance unit he called an *astronomical unit* or AU. One AU represents the length of Earth's semimajor axis; that is, the Earth's average distance from the Sun. Today, as mentioned earlier, this distance is known to be about 93,000,000 miles or 149,600,000 kilometers.

By 1616 Kepler had determined a relationship between the average distance of a planet from the Sun to its period of revolution. The period of time he used was the time it took planets to make one circuit of the sky with respect to the stars. This time interval is the planet's *sidereal period.*

The third law of Kepler is known as the *harmonic law*. Simply stated, the square of the sidereal period (P) is directly proportional to the cube of the semimajor axis (a) of the planet's orbit.

This law is expressed in a symbolic fashion in Equation 6-1.

$$P^2 = ka^3 \hspace{4cm} \text{Equation 6-1}$$

The k is a constant of proportionality in this expression. If the sidereal period (P) is expressed in years and the average distance (a) is expressed in astronomical units (AUs), then the constant of proportionality, k, in Equation 6-1 equals 1. With this judicious choice of units of P and a, Equation 6-1 may be written in the form shown in Equation 6-2.

$$P^2 = a^3 \hspace{4cm} \text{Equation 6-2}$$

Kepler's third law may be thought of as the "yardstick" of the Solar System. In other words, if the period of time it takes a planet to complete one revolution around the Sun (in years) can be measured, then it is possible to determine its distance on the average in AUs. Or, if a planet's average distance is known, then it is possible to calculate the planet's period of revolution around the Sun with respect to the stars.

Distances and periods of revolution that are calculated using Kepler's third law are *relative* distances and periods. That is, they are determined in terms of Earth's distance from the Sun and its period of revolution. This law is very useful in determining distances and periods of revolution of newly discovered objects in the Solar System, such as asteroids and comets!

When using *TheSky*, you will find the physical and orbital data for the planets very useful. To take advantage of this information, make sure the Student Edition of *TheSky* CD is in the your CD-ROM drive. Clicking on the Multimedia tab in the Object Information window gives you access to all the physical and orbital data on the planets. These data are stored in a text file named planetname.txt. "Planetname" is just a generic name for the planet's name in *TheSky's* database.

Configurations of the Planets

Our Solar System, as stated earlier, is a flat, counterclockwise system. Because the entire Solar System is in motion, it is not hard to imagine that the planets can line up with each other as they orbit the Sun. When these alignments occur, they are called *planetary configurations.* Configurations of planets are important when you are trying to understand their motions in the sky or their appearance in a telescope.

Knowing the configuration of a planet actually allows you to locate where the planet is in its orbit with respect to the Sun and Earth. An easy method of determining the configuration of a planet is to measure an angle. This angle is defined as a planet's *elongation.* It is the angle measured between a line drawn from Earth to the Sun and a line drawn from the Earth to the planet. The measurement of elongation is shown in Figure 6-5.

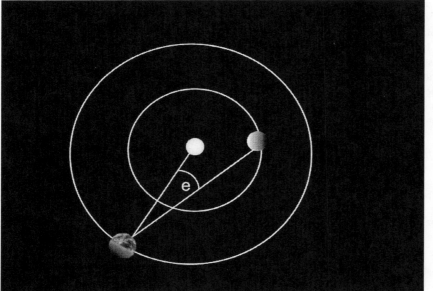

Figure 6-5 Measuring elongation

In the sky, this angle is measured from an Earth–Sun line to an Earth–planet line along the ecliptic. The elongation of planets varies as they revolve around the Sun. Because an inferior planet (closer) moves around the Sun faster than Earth, it goes through a variety of configurations more quickly than a superior (farther) planet does. The configurations of an inferior planet are displayed in Figure 6-6.

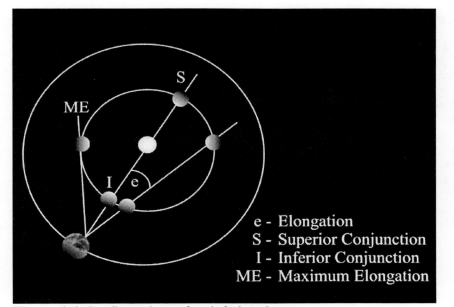

e - Elongation
S - Superior Conjunction
I - Inferior Conjunction
ME - Maximum Elongation

Figure 6-6 Configurations of an inferior planet

As the inferior planet moves around the Sun, its elongation constantly changes. In Figure 6-6 the letter *e* represents the elongation angle. The letters *ME* represent the maximum elongation of an inferior planet. When inferior planets are observed in Earth's sky, they never appear too far (in angle) from the Sun.

Sometimes inferior planets appear east of the Sun in Earth's sky, at other times they appear west of it. When an inferior planet is east of the Sun, it appears to rise after the Sun does and appears to set after the Sun does. It is visible during early evening twilight hours and is often referred to as an "evening star."

When an inferior planet is west of the Sun, it appears to rise before the Sun and appears to set before the Sun. It is visible during the early morning twilight hours and is often referred to as a "morning star."

People are often confused about how to determine directions from the Sun when it comes to measuring elongations in the sky. To put it simply—which way is east or west? To make sense out of all this, you must first try to properly orient yourself on the Sun.

To do this, it is helpful to go outside and observe the Sun's position at midday. The Sun's location in the sky, of course, is due south on the local celestial meridian. While standing outside facing south, imagine seeing a planet to the left or right of the Sun. If the planet is left of the Sun in the sky, then it is east of the Sun. If the planet is right of the Sun in the sky, then it is west of the Sun. No matter where the Sun is located on the ecliptic, whenever a planet is to the left of it, it is in eastern elongation. Whenever the planet is to the right of the Sun, it is in western elongation.

Figure 6-7 graphically illustrates the direction that a planet may be with respect to the Sun in Earth's sky. When you as the observer face south, notice that east is to your left and west is to your right as in this figure.

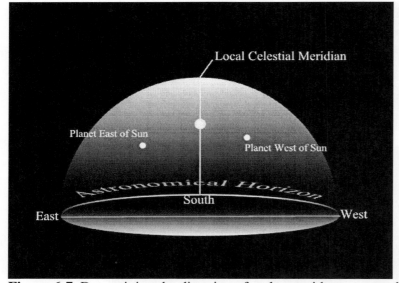

Figure 6-7 Determining the direction of a planet with respect to the Sun

This orientation is true whenever you use *TheSky*. Basically, if a planet is left of the Sun in the sky window, then it is east of the Sun and if it is right of the Sun in the sky window, then it is west of the Sun. Figure 6-8 depicts this relationship.

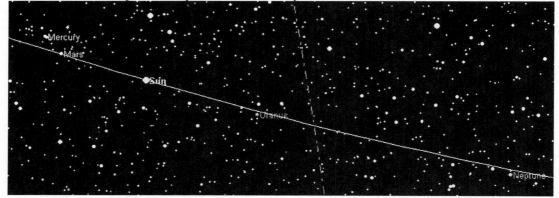

Figure 6-8 Locating planets in *TheSky*

The elongations of inferior planets have values that range from 0° (in line with the Sun) to some maximum value. The maximum elongation for an inferior planet is defined as the planet's *greatest eastern* or *western elongation*. For the planet Mercury, this angle is about 26° and for Venus it is close to 48°. You must realize that these angles are for observations made under ideal conditions. The orbits of the planets do not lie exactly in the same plane with each other but are slightly tilted, as you have seen in previous multimedia files.

The multimedia file named VenusGEE.sky illustrates a 3D view of the planets Venus and Earth with the Sun. Venus is located at greatest eastern elongation. Exercise U demonstrates the position of Earth, Venus, and the Sun when Venus is at its greatest angular separation east of the Sun. Notice that Venus appears to the left of the Sun as viewed from Earth's position in the Solar System.

Exercise U: Observing Venus at Greatest Eastern Elongation

1. Open the multimedia file named VenusGEE.sky. Using the scroll bar at the right edge of the window, observe the configuration from different vantage points above the Solar System.

2. Close the file. It is not necessary to save this multimedia file.

The multimedia file named MercuryGWE.sky is also a 3D window illustrating the planets Mercury and Earth with the Sun. Mercury is located at greatest western elongation. Exercise V demonstrates the position of Mercury, Earth, and the Sun when Mercury is at its greatest angular separation west of the Sun. Notice that Mercury appears to the right of the Sun as viewed from Earth's position in the Solar System.

Exercise V: Observing Mercury at Greatest Western Elongation

1. Open the multimedia file named MercuryGWE.sky. Using the scroll bar at the right edge of the window, observe the configuration from different vantage points above the Solar System.

2. Close the file. It is not necessary to save this multimedia file.

The greatest eastern and western elongations are the two extreme angles measured for an inferior planet. So any location between these and any alignment with the Sun is usually described as the planet being in either eastern or western elongation.

As an inferior planet orbits the Sun, at two locations in its orbit the elongation is 0°. That is, the planet aligns itself with the Sun as viewed from Earth. One location is where the planet is more distant than the Sun and the other is where the planet is between Earth and the Sun. These locations are labeled S and I, respectively, in Figure 6-6.

When the planet is located at S, it is aligned with the Sun as viewed from Earth, and its elongation is 0°. This configuration is defined as *superior conjunction* of an inferior planet. The multimedia file named VenusSupC.sky gives a 3D representation of this configuration.

When the planet is located at I, it is aligned with the Sun again as viewed from Earth, and its elongation is again 0°. This configuration is defined as *inferior conjunction* of an inferior planet. The multimedia file named VenusInfC.sky is also a 3D representation that illustrates this configuration.

At either location, the angular separation from the Sun is 0°. In other words, an inferior planet would have the same right ascension as the Sun and appear to be very close to the Sun in Earth's sky. Exercise W illustrates Venus at the configuration of superior conjunction. Exercise W illustrates Venus at the configuration of inferior conjunction. In both of these exercises, from Earth's position in the Solar System, Venus appears to be in the direction of the Sun.

Exercise W: Observing Venus at Superior Conjunction

1. Open the multimedia file named VenusSupC.sky.

2. Using the scroll bar at the right edge of the window, observe this configuration from different vantage points above the Solar System.

3. Click the "3D Solar System Mode" button, and notice that Venus and the Sun appear in Earth's sky in the southwest. Notice how close they are together in the sky. Click the Day Sky mode Button and observe them.

4. Close the file. It is not necessary to save this multimedia file.

Exercise X: Observing Venus at Inferior Conjunction

1. Open the multimedia file named VenusInfC.sky. Using the scroll bar at the right edge of the window, observe the configuration from different vantage points above the Solar System.

2. Click the "3D Solar System Mode" button, and notice that Venus and the Sun appear in Earth's sky in the west. Notice that they are not as close as they were in the previous exercise. It may not be apparent in this sky window that Venus and the Sun have about the same right ascension. Click on the equatorial grid and estimate the right ascension. They are very close. Click the Day Sky mode Button and observe them.

3. Close the file. It is not necessary to save this multimedia file.

The elongations for Mercury and Venus are shown in Figure 6-9. Also included in the figure are approximate times when these planets appear to rise and set.

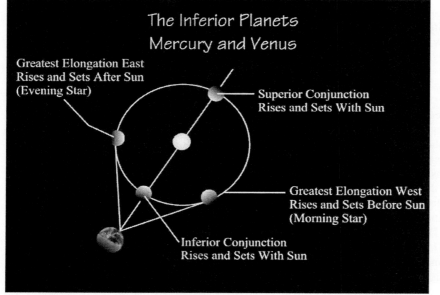

Figure 6-9 Elongation of Mercury and Venus

Even when Mercury is at either of its greatest elongations, it is usually seen in the twilight sky. The reason for this is that it appears to rise and set approximately one to two hours before and after the Sun does, respectively. Venus, in contrast, is often seen in a totally dark sky. In fact, it's the third brightest object in the sky. When it is at either of its greatest elongations, Venus appears to rise and set approximately three hours before or after the Sun does, respectively.

As superior planets orbit the Sun their orbital positions change, but at a much slower rate than the inferior planets. Therefore their elongations vary as well when they are viewed from Earth. However, because a superior planet's orbit is greater than that of Earth, the elongation can be much greater than that of an inferior planet.

Elongations for superior planets range in values from 0° to 180°. When a superior planet's elongation is 0°, it too has the same right ascension as the Sun. This configuration is defined as *conjunction*. The multimedia file named JupiterConj.sky illustrates a 3D representation of this configuration. It depicts Jupiter's position in the Solar System on May 8, 2000, relative to Earth and the Sun.

Exercise Y illustrates the planets Jupiter and Earth with the Sun in a 3D representation of Jupiter in conjunction. You will notice that Jupiter is in the same direction of the Sun as viewed from Earth's position in the Solar System.

Exercise Y: Observing Jupiter in Conjunction

1. Open the multimedia file named JupiterConj.sky. Using the scroll bar at the right edge of the window, observe the configuration from different vantage points above the Solar System.

2. Click the "3D Solar System Mode" button, and notice that Jupiter and the Sun are seen below the northeastern horizon. Notice also how close together they appear in the sky.

3. Click on the Sun and note its equatorial coordinates. Now click on Jupiter and note its equatorial coordinates. Are they the same as those of the Sun? _____. You can find the angular separation between Jupiter and the Sun in the Object Information window. What is Jupiter's angular separation from the Sun (you must click on the Sun first then Jupiter)? _____.

4. Close the file. It is not necessary to save this multimedia file.

As a superior planet continues to move counterclockwise around the Sun from conjunction, its elongation changes. All the while the planet appears to be left of the Sun in the sky, or in eastern elongation. Eventually, it reaches a location in its orbit that is 90° from the Sun. This configuration is defined as a *quadrature*. When a superior planet's position is 90° east of the Sun, its configuration is defined as *eastern quadrature*. The multimedia file named JupiterEQ.sky portrays a 3D representation of Jupiter at the time of eastern quadrature in January 2000. Exercise Z illustrates the planets Jupiter and Earth with the Sun in a 3D view of Jupiter in the configuration of eastern quadrature. Notice that Jupiter is 90° to the left of the Sun as viewed from Earth's position in the Solar System.

Exercise Z: Observing Jupiter at Eastern Quadrature

1. Open the multimedia file named JupiterEQ.sky. Using the scroll bar at the right edge of the window, observe the configuration from different vantage points above the Solar System.

2. Click the "3D Solar System Mode" button, and notice that the Sun is close to the local celestial meridian. Find Jupiter and center it. Notice that Jupiter has just risen in the east.

3. Find the Sun and note its equatorial coordinates. Now find Jupiter and center it in the sky window. Note the equatorial coordinates of Jupiter. Are they about 6^H greater (eastward) than that of the Sun? _____. In the Object Information window you can find the angular separation of Jupiter from the Sun. Click on the Sun first, then on Jupiter. Is it about 90°? _____.

4. Close the file. It is not necessary to save this multimedia file.

As a superior planet continues its motion in orbit, it eventually reaches a location where it is exactly 180° from the Sun. This configuration is defined as *opposition*. The multimedia file named JupiterOpp.sky displays the planets Jupiter and Earth and the Sun at the time of opposition. You will notice that the Earth is between Jupiter and the Sun; Jupiter is exactly 180° from the Sun as viewed from Earth.

Whenever a superior planet is at opposition, it appears to rise at the time the Sun appears to set. It is visible all night long. During the interval of time that a superior planet approaches opposition, it also appears to retrograde in Earth's sky. Exercise AA illustrates the planets Jupiter and Earth with the Sun in a 3D representation at the time of opposition.

Exercise AA: Observing Jupiter at Opposition

1. Open the multimedia file named JupiterOpp.sky. Using the scroll bar at the right edge of the window, observe the configuration from different vantage points above the Solar System.

2. Click the "3D Solar System Mode" button, and you notice that the Sun appears to be setting in the west-southwest. Find Jupiter and center. Did you notice where it is with respect to the Sun? _____.

3. Find the Sun and note its equatorial coordinates. Now find Jupiter and center it in the sky window. Note the equatorial coordinates of Jupiter. Are they about 12^H greater (eastward) than that of the Sun's? _____. In the Object Information window you can find the angular separation between Jupiter and the Sun. Is it about 180°? _____.

4. Close the file. It is not necessary to save this multimedia file.

As the superior planet continues its motion around the Sun, it eventually ends up at a location in its orbit 90° from opposition. It is also 90° from the Sun once again. This time, however, it is now 90° west of the Sun instead of east. When a superior planet is 90° west of the Sun, its configuration is defined as *western quadrature*.

The multimedia file named JupiterWQ.sky is a 3D representation of Jupiter, Earth, and the Sun at the time of western quadrature. Exercise BB illustrates the planets Jupiter and Earth with the Sun at the time when Jupiter is at western quadrature. You will notice that Jupiter is 90° to the right of the Sun as viewed from Earth's position in the Solar System.

The configurations of a superior planet are shown in Figure 6-10.

Exercise BB: Observing Jupiter at Western Quadrature

1. Open the multimedia file named JupiterWQ.sky. Using the scroll bar at the right edge of the window, observe the configuration from different vantage points above the Solar System.

2. Click the "3D Solar System Mode" button and you will notice that the Sun appears to be rising in the east and that Jupiter appears to be near the local celestial meridian.

3. Find the Sun and note its equatorial coordinates. Now find Jupiter and center it in the sky window. Note the equatorial coordinates of Jupiter. Are they about 6^H less (westward) than that of the Sun? _____. In the Object Information window you can find the angular separation of Jupiter from the Sun. Is it about 90°? _____.

4. Close the file. It is not necessary to save this multimedia file.

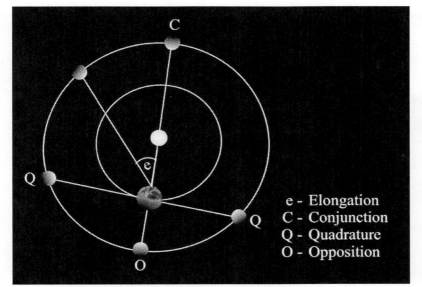

Figure 6-10 Configurations of a superior planet

Figure 6-11 shows the elongations of the superior planets as well as their times of visibility.

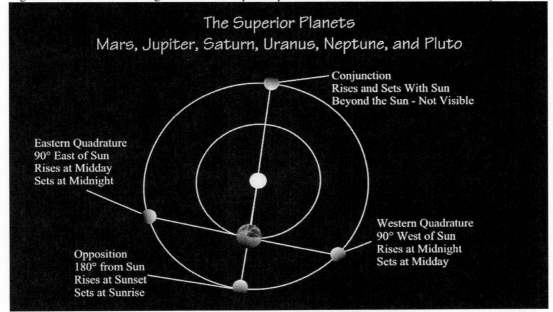

Figure 6-11 Elongation and times of visibility of a superior planet

Time intervals are also measured between planetary configurations. The interval of time between two successive occurrences of the same configuration is defined as the *synodic period* of a planet. This time interval is measurement of time with respect to the Sun, *not* the stars. It is quite different from the sidereal period of the planet. This period of time includes the motions of Earth and the planet.

Planetary Phases

As an inferior planet moves around the Sun, not only does its position change with respect to the Sun and Earth but the way sunlight strikes the planet also changes. When observing an inferior planet through telescopes from Earth, it is evident that Mercury and Venus go through phases like our Moon.

This phasing process, observed from Earth, is based on the synodic periods of the inferior planets. When Venus is located at greatest eastern or western elongation, it appears as a half circle or a quarter phase when viewed through a telescope. Figure 6-12 is *TheSky's* rendition of this phase.

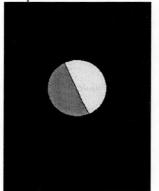

Figure 6-12 Phase of Venus at greatest eastern elongation

As Venus travels from greatest eastern elongation to greatest western elongation in its orbit, it goes through a series of phases. From its quarter phase, at greatest eastern elongation, it goes through many crescent phases that look similar to the image displayed in Figure 6-13.

During this time period Venus actually passes between Earth and the Sun through inferior conjunction. The closer Venus gets to the Sun, in angle in the sky, the more difficult it becomes to see. When Venus reaches inferior conjunction, it is aligned with the Sun and is not visible from Earth, because its dark hemisphere is facing Earth.

As Venus continues to move, from inferior conjunction toward the greatest western elongation, it again appears as a crescent phase when viewed from Earth. The image displayed in Figure 6-13 is Venus4.jpg from *TheSky's* image database. The Hubble Space Telescope took the image in Figure 6-13. Venus was in western elongation when this image was taken.

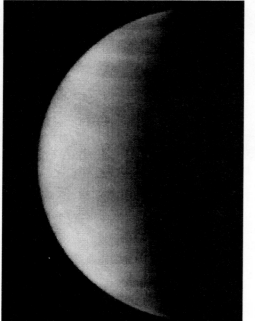

Figure 6-13 Crescent Venus in western elongation

The multimedia file VenusGEE_GWE.sky was rendered to display the phases of Venus as it travels from its greatest eastern elongation to its greatest western elongation. As Venus moves from greatest eastern elongation to inferior conjunction, its distance decreases as it approaches Earth. After passing inferior conjunction, its distance from Earth increases as it moves toward greatest western elongation.

During this period of time Venus is observed going through a series of phases from Earth. It goes from a quarter phase on the right edge through a series of crescent phases to a new phase when it reaches inferior conjunction. Venus and Earth are now at their closest distance to one another. The angular size of Venus also changes. Venus' angular size is at its greatest when in inferior conjunction.

From the new phase, it continues through a series of crescent phases. The crescent phases are now on the left edge of Venus, like the image in Figure 6-13. On reaching greatest western elongation, it appears as a quarter phase on the left edge of the planet. Figure 6-14 is *TheSky's* rendition of Venus' phase at greatest western elongation.

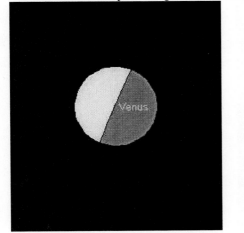

Figure 6-14 Phase of Venus at greatest western elongation

Exercise CC demonstrates the phases that Venus exhibits as it moves from greatest eastern elongation to greatest western elongation. In this sky window Venus appears to go through a series of crescent phases through the new phase.

As it moves from greatest eastern elongation to inferior conjunction, the crescent phase is on the right edge of the planet. As it moves from inferior conjunction to greatest western elongation, the crescent phase is on the left edge of the planet. It appears as a new phase when it is at inferior conjunction. The planet appears to get bigger in angular size as it approaches Earth (to inferior conjunction), and smaller as it moves away (from inferior conjunction) from Earth. This is seen in the multimedia file named VenusGEE_GWE.sky. The time period of this observation is from January 2001 to June 2001.

Exercise CC: Observing the Phases of Venus from Greatest Eastern to Greatest Western Elongation

1. Open the multimedia file named VenusGEE_GWE.sky. Click the ⏩ button and observe.

2. Clicking on the "3D Solar System Mode" button shows where Venus is with respect to the Sun and Earth.

3. Close the file. It is not necessary to save this multimedia file.

When Venus passes greatest western elongation in its orbit, its distance is actually farther from Earth than the Sun. As Venus travels from greatest western elongation to greatest eastern elongation in its orbit, it continues going through many phases. This time, however, the phases are known as gibbous phases.

A gibbous phase is larger than a quarter phase but smaller than a full phase. A gibbous phase looks like the image displayed in Figure 6-15. This image (Venus2.jpg) was taken from *TheSky's* image database.

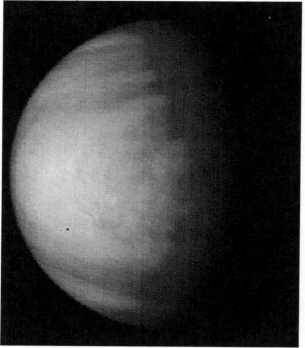

Figure 6-15 Gibbous Venus in western elongation

From its quarter phase, at greatest western elongation, more of the left edge of Venus becomes visible from Earth, as seen in Figure 6-15. Venus was in western elongation when this image was taken. The gibbous phase continues to get larger, or more of Venus becomes visible from Earth. During this time period Venus' distance from Earth continues to increase and the planet appears to get closer to the Sun in the sky (in angle). It becomes more difficult to see from Earth.

Midway through the time period between greatest western elongation and greatest eastern elongation, Venus passes the location that is in direct alignment with the Sun once again (superior conjunction). Venus is also not visible at this time. However, if it could be seen from Earth (perhaps during a solar eclipse), it would look like a full Moon phase. At superior conjunction its illuminated hemisphere is toward the Earth. Venus and Earth are now at their farthest distance from one another. The angular size of Venus is at its smallest when in superior conjunction.

As Venus continues to move, from superior conjunction toward the greatest eastern elongation, it again appears as a gibbous phase when viewed from Earth. During this time period, however, the right edge of Venus is illuminated instead of the left. It continues to go through a series of gibbous phases until it reaches greatest eastern elongation. Now it is a quarter phase once more.

The multimedia file named VenusGWE_GEE.sky was rendered to display the phases of Venus as it travels from its greatest western elongation to its greatest eastern elongation. As Venus moves from greatest western elongation to superior conjunction, its distance increases as it moves away from Earth. On reaching superior conjunction the distance between Venus and Earth is at its maximum distance and its angular size is the smallest. After passing superior conjunction, Venus' distance from Earth begins to decrease while it moves toward greatest eastern elongation.

During this period of time Venus is observed going through a series of phases from Earth. It goes from a quarter phase on the left edge through a series of gibbous phases and to a full phase when it reaches superior conjunction. Venus and Earth are now at their farthest distance from one another. Venus' angular size is at its smallest when at superior conjunction.

From the full phase, it continues through a series of gibbous phases. The gibbous phases are now on the right edge of Venus. On reaching greatest eastern elongation, it appears as a quarter phase on the right edge of the planet.

Exercise DD demonstrates the phases that Venus exhibits as it moves from the configuration of greatest western elongation to greatest eastern elongation. In this sky window Venus appears to move through a series of gibbous phases. As it moves from greatest western elongation to superior conjunction, the gibbous phase is on the left edge of the planet. As it moves from superior conjunction to greatest eastern elongation, the gibbous phase is on the right edge of the planet. It appears as a full phase when it is at superior conjunction. It appears to get smaller in angular size as it moves away from Earth (to superior conjunction) and larger as it approaches Earth (from superior conjunction). The time period for this observation is January 2003 to April 2004.

Exercise DD: Observing the Phases of Venus from Greatest Western to Greatest Eastern Elongation

1. Open the multimedia file named VenusGWE_GEE.sky. Click the [▶▶] button.

2. Clicking on the "3D Solar System Mode" button shows where Venus is with respect to the Sun and Earth.

3. Close the file. It is not necessary to save this multimedia file.

The multimedia file named VenusPhase.sky demonstrates all the phases of Venus during one revolution around the Sun. Exercise EE illustrates what is observed from Earth.

Exercise EE: Observing the Phases of Venus

1. Open the multimedia file named VenusPhase.sky. Click the [▶▶] button. Observe Venus as it goes through an entire set of phases through one synodic period. Notice too that its angular size also changes through this time period.

2. Can you estimate Venus' position in its orbit relative to Earth and the Sun at the beginning of this sky window? _____

3. Close the file. It is not necessary to save this multimedia file.

The multimedia file named VenusOrbit.sky displays a 3D representation of the sky multimedia file named VenusPhase.sky. Exercise FF demonstrates a 3D view of what was seen in the previous exercise.

Exercise FF: 3D Representation of VenusPhase.sky

1. Open the multimedia file named VenusOrbit.sky. Click the [▶▶] button. Observe Venus as it moves around the Sun through one synodic period.

2. Close the file. It is not necessary to save this multimedia file.

The fact that Venus goes through a complete set of phases is historically significant. In 1609 the Italian astronomer Galileo Galilei was the first person to report his observations of Venus through a small telescope. His hand-made telescope revealed that Venus exhibited phases like that of our Moon. From these observations, he concluded that the Sun was the center of our Solar System, instead of the Earth. His observations, he claimed, supported the heliocentric, or sun-centered, "universe" proposed by Nicolas Copernicus and Aristarchus of Samos.

In the case of a superior planet, however, the motion is much slower than that of Mercury or Venus. As a consequence, its change of position in the sky is less conspicuous than that of an inferior planet. The phases of a superior planet are *not as* extreme as an inferior planet. In a small telescope they almost always appear as a full phase. In larger instruments, however, their phases are subtle but more pronounced.

Phase information about the planets is found in the Object Information window. After locating a planet, simply click on the Observer's Log icon ( ) on the toolbar at the bottom of the Object Information window. It is found next to the printer icon on the toolbar as displayed in Figure 6-16.

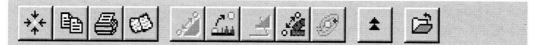

Figure 6-16 Observer's Log toolbar in the Object Information window

By clicking on the Multimedia tab in the Object Information window, you can obtain physical data about each of the planets. Figure 6-17 illustrates the text file location containing the physical data on Jupiter. It is the last file in the multimedia window index and is labeled as Jupiter.txt.

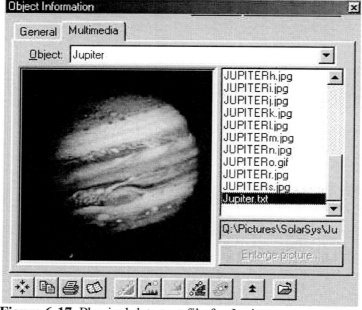

Figure 6-17 Physical data text file for Jupiter

Figure 6-18a and Figure 6-18b illustrate orbital and physical data contained in the Jupiter.txt file, respectively. *TheSky* database contains information on all the planets.

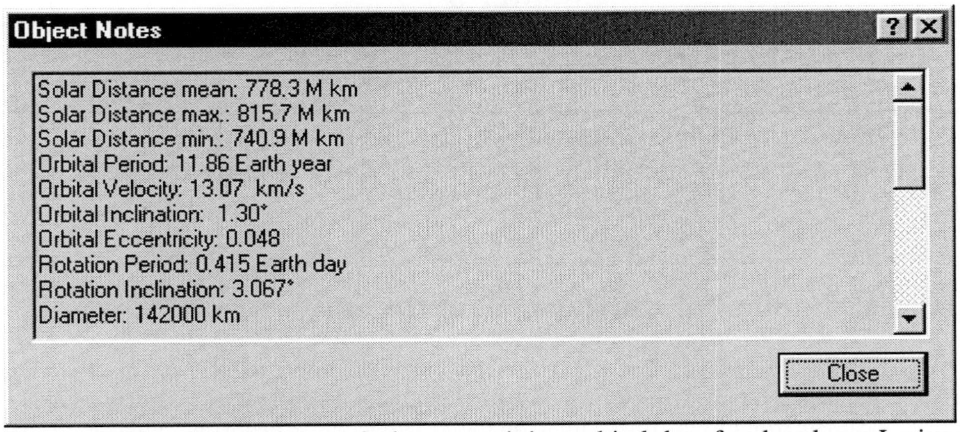

Figure 6-18a Object Notes window containing orbital data for the planet Jupiter

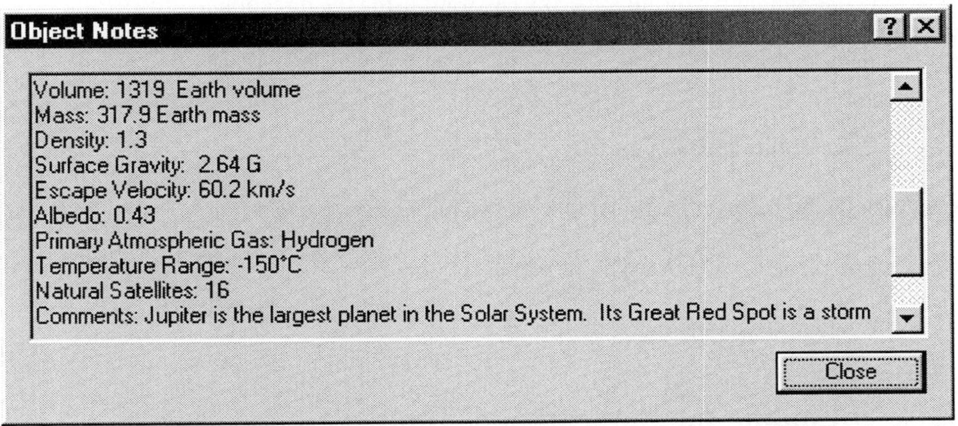

Figure 6-18b Object Notes window containing physical data on the Planet Jupiter

Chapter 6

TheSky Review Exercises

Start *TheSky*. You may use your computer's clock for the time in the first exercise.

TheSky Exercise 1: Finding Information About the Sun

1. Find the Sun in *TheSky*. Make sure that the Sun and planet labels are toggled on; use Filters in the <u>V</u>iew menu at the top of the sky window to select them if necessary. You may want to change Local Horizon Fill to "transparent" in the Reference <u>L</u>ine menu, as well.

2. After locating the Sun, move through the sky eastward using the directional arrows (blue) on the Orientation toolbar. Observe how the Sun appears to move through the stars. What color is this path in *TheSky*? _____ Remember that the color may be changed at any time.

3. What is the name of this apparent path? _____

4. While moving along the path in Question 2, did you notice any planets? _____

5. Were any of the planets along this path farther away than ±8°? _____

6. If the answer in Question 5 was no, why not? _____

TheSky Exercise 2: Finding the Elongation of Jupiter

Now set *TheSky* to the following:

Location: Golden, Colorado
Date: December 31, 2009
Time: 18:00 hours

1. In what constellation is the Sun currently located? _____

2. In what constellation is Jupiter currently located? _____

3. What are the right ascension and declination of Jupiter?

   R.A. = ___^{H} ___^{M} Declination = __ ___°___' (+, N; –, S)

4. What is its magnitude? _____

5. What is Jupiter's elongation (angular separation)? _____

6. Is it currently in eastern or western elongation? _____

TheSky Exercise 3: Finding the Elongation of Mars

1. In what constellation is Mars currently located? _____

2. What are the right ascension and declination of Mars?

   R.A. = ___^{H} ___^{M} Declination = __ ___° ___' (+, N; –, S)

3. What is its magnitude? _____

4. What is Mars' elongation? _____

5. Is it currently in eastern or western elongation? _____

TheSky Exercise 4: Finding the Elongation of Saturn

1. In what constellation is Saturn currently located? _____

2. What are the right ascension and declination of Saturn?

   R.A. = ___^{H} ___^{M} Declination = __ ___° ___' (+, N; –, S)

3. What is its magnitude? _____

4. What is Saturn's elongation? _____

5. Is it currently in eastern or western elongation? _____

TheSky Exercise 5: Finding the Elongation of Jupiter

Now, set *TheSky* to the following:
Location: Ball State Observatory
Date: December 31, 2009
Time: 18:00 hours

1. In what constellation is the Sun currently located? _____

2. In what constellation is Jupiter currently located? _____

3. What are the right ascension and declination of Jupiter?

   R.A. = ___^{H} ___^{M} Declination = __ ___° ___' (+, N; –, S)

4. What is its magnitude? _____

5. What is Jupiter's elongation (angular separation)? _____

6. Is it currently in eastern or western elongation? _____

TheSky Exercise 6: Finding the Elongation of Mars

1. In what constellation is Mars currently located? _____

2. What are the right ascension and declination of Mars?

    R.A. = ___^H ___^M   Declination = __ ___° ___′ (+, N; −, S)

3. What is its magnitude? _____

4. What is Mars' elongation? _____

5. Is it currently in eastern or western elongation? _____

TheSky Exercise 7: Finding the Elongation of Saturn

1. In what constellation is Saturn currently located? _____

2. What are the right ascension and declination of Saturn?

    R.A. = ___^H ___^M   Declination = __ ___° ___′ (+, N; −, S)

3. What is its magnitude? _____

4. What is Saturn's elongation? _____

5. Is it currently in eastern or western elongation? _____

TheSky Exercise 8: Comparing Measurements with *TheSky*

Compare Exercise 2 and Exercise 5:

1. Were there any differences in the observations from the two locations? _____

2. If so, what were they? _____
   _____

TheSky Exercise 9: Comparing Measurements with *TheSky*

Compare Exercise 3 and Exercise 6:

1. Were there any differences in the observations from the two locations? _____

2. If so, what were they? _____
   _____

TheSky Exercise 10: Comparing Measurements with *TheSky*

Compare Exercise 4 and Exercise 7:

1. Were there any differences in the observations from the two locations? _____

2. If so, what were they? _____
    _____

Chapter 6

TheSky Review Questions

Answer the following review questions using *TheSky* where necessary.

TheSky Review Question 1

1. What are the right ascension and declination of the point in the sky where the ecliptic and the celestial equator intersect in March each year? The Sun crosses this point from the Southern Hemisphere to Northern Hemisphere in the sky.

   R.A. = ___^{H} ___^{M} Declination = __ ___° ___′ (+, N; –, S)

2. What is the name given to this point in space? _____ _____

3. Against what constellation does the Sun appear to be projected in the sky when it is located at this point? _____

4. Now change the view from Earth looking toward the Sun, to that of the Sun looking toward the Earth. Use the blue directional buttons (*Hint*: Look 180° from Sun. Turn on the equatorial grid, if necessary, and move 12^{H} in right ascension eastward). Against what constellation would the Earth appear to be projected when viewed from the Sun on the date in Question 1? _____

5. What time of day would the constellation in Question 4 appear to be on your local celestial meridian? _____

TheSky Review Question 2

1. On what date does the Sun appear to be 23.5° North of the celestial equator? _____

2. What are the right ascension and declination of the point in Question 1?

   R.A. = ___^{H} ___^{M} Declination = __ ___° ___′

3. What is the name of this point in space? _____ _____

4. Against what constellation does the Sun appear to be projected on the date in Question 1?
   _____

5. Now change the view from Earth looking toward the Sun, to that of the Sun looking toward the Earth. Use the blue directional buttons again (*Hint*: Look 12^{H} in right ascension eastward). Against what constellation would Earth appear to be projected when viewed from the Sun on the date in Question 1? _____

6. What time of day would the constellation in Question 5 appear to be on your local celestial meridian? _____

TheSky Review Question 3

1. What are the right ascension and declination of the point where the ecliptic and celestial equator intersect in September each year? The Sun crosses this point from the Northern Hemisphere to the Southern Hemisphere in the sky?

    R.A. = ___^{H} ___^{M} Declination = __ ___° ___' (+, N; –, S)

2. What is the name given to this point in space? _____ _____

3. Against what constellation does the Sun appear to be projected when it is located at this point in the sky? _____

4. Now change the view from Earth looking toward the Sun, to that of the Sun looking toward the Earth (use the blue directional buttons again). Against what constellation does the Earth appear to be projected when viewed from the Sun on the date in Question 1?
    _____

5. What time of day would the constellation in Question 4 appear to be on your local celestial meridian? _____

TheSky Review Question 4

1. On what date does the Sun appear to be 23.5° south of the celestial equator?
    _____

2. What are the right ascension and declination of the point in Question 1?

    R.A. = ___^{H} ___^{M} Declination = __ ___° ___'

3. What is the name of this point in space? _____ _____

4. Against what constellation does the Sun appear to be projected on the date in Question 1?
    _____

5. Now change the view from Earth looking toward the Sun, to that of the Sun looking toward the Earth (use the blue directional buttons again). Against what constellation does the Earth appear to be projected when viewed from the Sun on the date in Question 1?
    _____

6. What time of day would the constellation in Question 5 appear to be on the local celestial meridian? _____

Chapter 7

Keeping Time in *TheSky*

One of the beauties of nature that men and women can truly enjoy is observing the heavens. The diurnal motions of the celestial bodies, changing of constellations with the seasons—the return of old "celestial friends"—are all there for us to observe and enjoy if we only take the time.

Because we inhabit a satellite of the Sun, and it in turn is moving, our planet has several motions. Partly because of these motions, many objects in the sky appear to move across our sky. Ever since the dawn of civilization people have become accustomed to thinking of Earth as being stationary and everything else in the sky as having motion. From our point of view, it does! But, looking at our system from a different frame of reference, things look quite different.

You can specify positions of celestial objects at any time in the sky by using the coordinate systems discussed in Chapter 5. However, if you wish to follow these objects as they change their positions in the sky you will need one additional parameter, and that is time.

Timekeeping

Over the centuries civilizations have devised ingenious ways to measure time. But time itself is a difficult concept to define. *You* try it! Define *time* without using terms that involve time itself—terms such as *between*, *during*, *from*, *to*, or *until*. All these terms imply the passage of time. In fact, physicists and astronomers have literally given up on defining time in any absolute sense. Instead, they use an *operational definition* for time. It is based on an isochronous process, which is something that happens over and over again.

One such process is to observe transits (or passages) of some reference object across your local celestial meridian each day. This reference object might be one of several objects. The Sun, of course, is the most obvious one to use. But, it could also be the Moon, a star, or simply a point in space. Every object in the sky appears to move at different rates in the sky, so we must choose an object that doesn't move very much over the duration of a day. Remember that a transit of some reference object across your local celestial meridian, that object's daily motion, is caused by Earth's rotation. Exercise A demonstrates this motion.

Exercise A: Observing Sun's Transit across the Local Celestial Meridian

1. Start *TheSky*.

2. Open the multimedia file named SunsTransit.sky.

3. Click on the [▶▶] button on the Time Skip toolbar.

4. Notice that the Sun appears to move from east to west across the celestial meridian (as well as everything else in the sky).

5. Close the file. It is not necessary to save this multimedia file.

Earth's revolution around the Sun makes the Sun appear to move 1° eastward (per day) along the ecliptic relative to the background stars. This motion is easily observed and can be precisely measured. This long-term motion of the Sun is demonstrated in the multimedia file named SunsAppMotion.sky. Exercise B illustrates the Sun's eastward motion against the background stars.

Exercise B: Measuring the Sun's Motion in *TheSky*

1. Start *TheSky*.

2. Open the multimedia file named SunsAppMotion.sky

3. Click on the [▶] button on the Time Skip toolbar.

4. Notice that the Sun moves in this window from close to the bright star α Leonis to a point in space that is approximately 1° east. You can measure how far the Sun moves in the sky. First, click on the Sun before using the Step Forward button, and then close the Object Information window. Next click on [▶] button again and then on the Sun once more. To find out how much the Sun moves, look for the angular separation information in the Object Information window. It is located near the bottom of the window.
    What is its value? ____°____'____". This is how far the Sun has moved relative to the background stars (its long-term motion).

5. To observe this motion for a period of several days, click the [▲] button then click the [▶▶] button, and finally click the [■] button.

6. Close the file. It is not necessary to save this multimedia file.

    The Moon, in contrast, revolves around us and moves approximately its own angular diameter (½°) per hour. This eastward motion of the Moon amounts to an angle of about 12° per day with respect to the Sun and 13° with respect to the stars. The motion of the Moon can be easily seen in the multimedia file named MoonsEMotion.sky. Exercise C illustrates the Moon's eastward motion projected against the distant background stars.

Exercise C: Measuring the Moon's Motion in *TheSky*

1. Start *TheSky*.

2. Open the multimedia file named MoonsEMotion.sky.

3. Click on the [▶] button on the Time Skip toolbar.

4. Notice that the Moon moves from near the planet Saturn to a point in space near the bright star α Tauri (Aldebaran). The motion results in an angle of approximately 13° eastward.
    You can verify this too. Click on the Moon before using the [▶] button then close the Object Information window. Next click on the [▶] button again and then on the Moon. To find out how far the Moon moves, find the angular separation near the bottom of the Object Information window like you did in Exercise A.
    What is its value? ____°____'____". This is how far the Moon has moved relative to the background stars (its long-term motion).

5. To observe this motion for a period of several days, click [▲] button then [▶▶] button, and Finally click the [■] button.

6. Close the file. It is not necessary to save this multimedia file.

When using the Moon as a reference object, however, its eastward motion presents us with some minor problems in timekeeping. It moves farther eastward, in angle, through the sky than any other object.

If we choose a star, in contrast, then the question becomes, Which star should we use as our reference object? You might think this isn't a problem, but two people may not agree on the same star to use as the reference object.

To eliminate confusion in choosing a particular star, astronomers have selected a well-defined point in space. Its position in the sky and its motion is well known. The reference object astronomers use to keep track of time by the stars is the vernal equinox. Figure 7-1 illustrates where the vernal equinox is located in space. It is at the intersection of the ecliptic and celestial equator where the Sun appears to cross the celestial equator in March. Its location, as shown in Figure 7-1, is directly on the local celestial meridian.

Figure 7-1 Location of the vernal equinox

We don't ignore the fact that the Sun is probably the most obvious object to use for timekeeping. In fact, the time interval between two successive passages of the center of the Sun's disk across your celestial meridian defines an *apparent solar day*. Remember that the Sun only appears to move east to west across the sky! This is caused by Earth's rotation and has a period of time equal to $24^H 00^M 00^S$.

When Earth's motion around the Sun is factored in, stars and the vernal equinox appear to move westward ~1° per day with respect to the Sun. Astronomers define the time interval between two successive passages of the vernal equinox across an observer's local celestial meridian as a *sidereal day*. This period of time is only $23^H 56^M 4.09^S$. A sidereal day is approximately 4 minutes shorter than a solar day.

The difference in time between a solar day and a sidereal day is actually caused by Earth's rotation and its orbital motion around the Sun. The sidereal day reflects the *true* rotation period of Earth. This is the actual length of time it takes Earth to rotate once on its axis. The solar day, in contrast, is a time interval that includes both the rotation and revolution of Earth. Each day as Earth rotates once on its axis it moves approximately 1° in its orbit. This 1° difference amounts to a difference of 4^M in time. Review the angle-to-time conversion discussed in Chapter 5.

Local Time

Timekeeping involves measuring an angle between some reference object in the sky and a specific celestial meridian. In fact, it can be measured from any celestial meridian at any location on Earth. The angle itself is measured westward along the celestial equator to the hour circle of the reference object. If you recall, an hour circle is a great circle on the celestial sphere passing through the celestial poles and an object on the celestial sphere. The angle is measured in hours and minutes of time, not in degrees and minutes of angle. This angle is appropriately defined as the *hour angle* of the reference object.

Anyone who can tell time actually measures an hour angle. That is, when you look at a watch or a clock on the wall, you are measuring the hour angle of the Sun. This is done from a particular celestial meridian at some location on Earth. Thus, the terms *A.M.* and *P.M.* mean *ante* (before) and *post* (past) meridian, respectively.

For practical reasons, those time systems based on the Sun use the hour angle of the Sun relative to an observer's celestial meridian and *add* 12 hours. This procedure is used so that the calendar date changes at midnight instead of at noon each day. Time kept by a local celestial meridian is defined as *local solar time* or simply *local time*. It is necessary to keep track of local time for events such as the apparent sunrise and sunset or the transits of celestial objects.

Standard Time

The system of standard time was established because observers at different longitudes on Earth do not have the same reference line (celestial meridian) in the sky. It is more convenient to standardize time at several locations on Earth rather than keeping track of their own local time. Cities separated by only a few hundred miles could keep the same time on their clocks. In the United States, during the 19th century, railroad companies were instrumental in implementing standard time that allowed train schedules to be put into operation within certain regions of the country.

Standard time is time kept by an observer located at a "standard longitude" on the surface of Earth. A standard longitude is a longitude circle that is a multiple of 15° in angle (30° E, 15° E, 0°, 15° W, 30° W). The hour angle of the Sun kept at one of these longitudes is defined as *standard* or "zone" *time*. These regions span a zone 15° wide on the surface of the Earth and are centered on a longitude circle that is a multiple of 15°.

There are four time zones across the United States. The Eastern Time Zone is centered on the 75th longitude circle west of Greenwich, England, and is very close to Philadelphia, Pennsylvania. The Central Time Zone is centered on the 90th longitude circle west of Greenwich and is very close to Memphis, Tennessee. The Mountain Time Zone is centered on the 105th longitude circle west of Greenwich and is very close to Denver, Colorado. The Pacific Time Zone is centered on the 120th longitude circle west of Greenwich and is very close to Vandenberg Air Force Base, California. Although observers throughout the zone have different local times, they do keep the same zone or standard time on their clocks. Figure 7-2 depicts the four time zones across the United States. The authors gratefully acknowledge the source of this map as a courtesy of www.theodora.com/maps and it is used with the permission of that Web site.

Figure 7-2 Standard time zones across the United States

One thing should be noted concerning Standard Time. Local events such as the apparent rising or setting of the Sun, Moon, and the planets do not occur at the same instant across the zone. Observers located on the eastern edge of the zone will observe events such as the Sun's apparent rising before an observer located at the center of the zone. Whereas, observers located on the western edge of the zone will observe events such as the Sun's apparent rising after an observer located at the center of the zone. Only their watch times will agree!

Local Time, in contrast, is the time that is kept at one's own location on Earth. It measures the hour angle of the apparent Sun from your local celestial meridian. On a sundial, this is the time that is measured with the gnomon, the marker. The shadow of the gnomon is cast on a plate with hours of time marked on it. It is necessary to keep track of this time so that events such as the apparent rising, transit, and setting of objects visible from your location can be determined. Figure 7-3 displays a typical sundial. See the Sun's shadow just to the left of the gnomon. The local time is approximately 10:00 AM.

Figure 7-3 A typical sundial
(Image courtesy of author)

Corrections in Time

TheSky computes local time for events that are visible from your location. In order to compute the local time of an astronomical event yourself, you must know the longitude of the your location to arcminute precision. The local time is then calculated by either *adding* (if west of the standard longitude) or by *subtracting* (if east of the standard longitude) a time correction to the standard longitude. This amounts to determining the difference in angle between your longitude circle and the standard longitude circle. The difference in time is calculated from the difference in angles and applied to the standard time as previously described.

An observer at the Ball State Observatory, longitude = 85° 21′ 38″, is exactly 10° 21′ 38″ west of the 75th longitude circle, which is the central longitude for the Eastern Standard Time zone. This angular difference amounts to a difference of about $41^M 26^S$ in time. This difference in time is then added to the "celestial event" time that occurs at the 75th longitude. In other words, "local noon" occurs at the Ball State Observatory approximately 12:41 P.M. EST every day.

The reason that it is necessary to keep track of local time is to determine precisely when astronomical objects such as the Sun, Moon, planets, appear to rise, transit your local celestial meridian, and set each day. This consequently determines a time when astronomical twilight begins or ends at a particular location on Earth. Astronomers define *astronomical twilight* as the moment when the center of the Sun's apparent disk is 18° below the astronomical horizon. Astronomers have adopted this definition to determine the instant when the sky is totally dark, unlike civil twilight, which is when the Sun's apparent disk is only 6° below the horizon. Civil twilight is what most states in the United States use to establish a time for turning car headlights on and off.

Universal Time

The last time system involving the Sun that we need to address is the one that keeps track of time for the world. This time system is kept at the prime meridian (0° longitude) in Greenwich, England. It used to be appropriately called Greenwich Mean Time but today is now known as Universal Time (UT). All clocks in the world are synchronized with the "world" clock located at the Royal Naval Observatory in Greenwich, England.

Events such as elongation of planets, lunar phases, lunar eclipses, solar eclipses, or the beginning of the seasons have their times determined in universal time. These times are then later converted to the observer's own standard time. Only events that involve apparent rising, transits, or setting of astronomical objects actually need corrections made to local time.

Events involving phenomena of the Sun, Moon, and planets are found in many resources. One of the most popular resources is *The Astronomical Almanac* that is issued annually by the United States Naval Observatory by direction of the Secretary of the Navy and under the authority of Congress. The U.S. Government Printing Office in Washington, D.C. publishes this almanac. Another excellent resource, for the amateur astronomer, is the *Observer's Handbook*. This resource is also issued annually by the Royal Astronomical Society of Canada. It is printed in Canada by the University of Toronto Press.

Astronomers use universal time extensively rather than using local or zone time. It makes sense to do so, because observations are made from observatories located in remote places on Earth. All observatory clocks worldwide are synchronized to universal time.

Sidereal Time

The last time system, and perhaps the most important one to astronomers, is sidereal time. *Sidereal Time* is defined as the *hour angle* of the vernal equinox. Why is sidereal time so important? Sidereal time is time kept by the stars.

When planning observing sessions, it is very important to know the sidereal time at your location on Earth. Knowing the sidereal time assists you in determining what is above the horizon at any location and consequently what is visible at that time! To accomplish this, you must determine the hour angle of the vernal equinox at the time of your planned observation.

Because sidereal time is defined as the hour angle of the vernal equinox, you must determine where the vernal equinox is located with respect to your local celestial meridian. The hour angle is measured westward, from the local celestial meridian along the celestial equator, to the hour circle of the vernal equinox.

Therefore, if you measure the angular distance westward along the celestial equator from your local celestial meridian to the hour circle of the vernal equinox, by definition you get the sidereal time at your location. Measurement of the hour angle of the vernal equinox is illustrated in Figure 7-4. The sidereal time in this figure is approximately 01:00, which means the vernal equinox is about one hour west of the local celestial meridian.

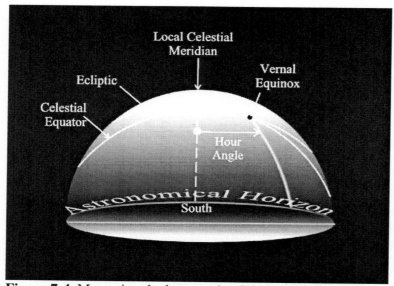

Figure 7-4 Measuring the hour angle of the Vernal Equinox

Measuring the hour angle of the vernal equinox from one's location on Earth essentially amounts to determining what the right ascension of objects are on your local celestial meridian. Remember, the coordinate of right ascension is the angle measured eastward from the vernal equinox (R.A. = 0^H) along the celestial equator to the hour circle of any celestial object. If the right ascension on your local celestial meridian is known, then the sidereal time is known!

Longitudinal Differences in Time

Exercises D, E, F, G and H are designed to demonstrate how *TheSky* is used to determine when an object, like a star, transits the local celestial meridian. They have been designed to confirm longitudinal effects of the observer's location on Earth. These effects will be observed when an object's transit time across the local celestial meridian is determined. To illustrate that there really is a difference in time, *TheSky* must be set *exactly* as in these examples.

Exercise D: Observing Longitudinal Differences in Time

1. Start *TheSky* and set as follows:

    Location: Ball State Observatory
    Ele<u>v</u>ation: 322 meters
    Date: August 25, 2002
    Time: 21:00   Time <u>Z</u>one: +5.00
    Daylight savings adjustment <u>o</u>ption: "Not Observed"
    Orientation: <u>M</u>ove To: R.A.: 18:56:32.1   Declination: 40° 11′ 8.7″ <u>N</u>orth
    Current Epoch   Orientation: Zoom <u>T</u>o: 50° field of view

    Turn the Equatorial Grid off and, and turn on the Horizon Grid.
    Make sure that the Bayer letters are displayed in the sky window.

    Find α Lyrae and determine its apparent transit time. To do this, click on α Lyrae and then click on the Rise, Set, and Transit button at the bottom of the Object Information window.

    Proper name: _____

Transit time: _____:_____

R.A. (2002): _____ Declination (2002): _____

Altitude: _____ Azimuth: _____

2. Change the setup as follows:

Location: Philadelphia, PA
Elevation: 100 meters
Date: August 25, 2002
Time: 21:00   Time Zone = +5.00
Daylight savings adjustment option: "Not Observed"
Orientation: Move To: R.A. = 18:56:32.1   Declination: 40° 11′ 8.7″ North
Current Epoch   Orientation: Zoom To: 50° field of view

Locate α Lyrae. Determine its apparent transit time. To do this, click on α Lyrae and then click on the Rise, Set, and Transit button at the bottom of the Object Information window.

Transit time: _____:_____

R.A. (2002): _____ Declination (2002): _____

Altitude: _____ Azimuth: _____

Note the differences between the results of Part 2 and Part 1, and record them in the following spaces:

Difference in transit times: _____: _____

Were there any differences in the altitude and azimuth? _____. What were they?

Difference in altitudes: _____ Difference in azimuths: _____

Why do the results for Part 2 and Part 1 differ? _____
_____

3. Change the setup as follows:

Longitude: 75° 00′ 00″
Elevation: 100 meters
Latitude: 40° 11′ 00″ (Same as Ball State Observatory)
Date: August 25, 2002
Time: 21:00   Time Zone: +5
Daylight savings adjustment option: "Not Observed"
Orientation: Move To: R.A.: 18:56:32.1   Declination: 40° 11′ 8.7″ North
Current Epoch   Orientation: Zoom To: 50° field of view

Locate α Lyrae. Determine its apparent transit time. To do this, click on α Lyrae and then click on the Rise, Set, and Transit button at the bottom of the Object Information window.

Transit time: _____:_____

R.A. (2002): _____ Declination (2002): _____

Altitude: _____ Azimuth: _____

Note the differences between the results of Part 3 and Part 1, and record them in the following spaces:

Difference in transit times: _____:_____

Were there any differences in the altitude and azimuth? _____. What were they?

Difference in altitudes: _____ Difference in azimuths: _____

Why do the results of Part 3 and Part 1 differ? _____
_____

4. Close the file and do not save, or continue with Exercise E.

Exercise E: Observing Longitudinal Differences in Time

1. Reset *TheSky* as follows:

Location: Ball State Observatory
Elevation: 322 meters
Date: April 20, 2010
Time: 05:20   Time Zone: +5.00
Daylight savings adjustment option: "Not Observed"
Orientation: Move To: R.A.: 20:45:30.0   Declination: 45° 30′ 15.0″ North
Current Epoch   Orientation: Zoom To: 50° field of view

Turn the Equatorial Grid off, and turn on the Horizon Grid.
Make sure that the Bayer letters are displayed in the sky window.

Locate α Cygni. Find its apparent rising, transit, and setting times.

Proper name: _____

Rising time: _____:_____ Transit time: _____:_____ Setting time: _____:_____

R.A. (2010): _____ Declination (2010): _____

Altitude: _____ Azimuth: _____

2. Change the setup as follows:

Location: Philadelphia, PA
Elevation: 100 meters
Date: April 20, 2010
Time: 05:20   Time Zone = +5.00
Daylight savings adjustment option: "Not Observed"
Orientation: Move To: R. A.: 20:45:30.0   Declination: 45° 30′ 15.0″ North
Current Epoch   Orientation: Zoom To: 50° field of view

Locate α Cygni. Find its apparent rising, transit, and setting times.

Rising time: _____:_____ Transit time: _____:_____ Setting time: _____:_____

R.A. (2010): _____ Declination (2010): _____

Altitude: _____ Azimuth: _____

Note the differences between the results of Part 2 and Part 1, and record them in the following spaces:

Differences in rising times: _____ Differences in transit times: _____

Differences in setting times: _____

Were there any differences in the altitude and azimuth? _____. What were they?

Difference in altitudes: _____ Difference in azimuths: _____

Why do the results between Part 2 and Part 1 differ? _____
_____

3. Change the setup as follows:

Longitude: 75° 00′ 00″
Elevation: 100 meters
Latitude: 40° 11′ 59″ (Same as Ball State Observatory)
Date: April 20, 2010
Time: 05:20   Time Zone: +5
Daylight savings adjustment option: "Not Observed"
Orientation: Move To: R. A.: 20:45:30.0   Declination: 45° 30′ 15. 0″ North
Current Epoch   Orientation: Zoom To: 50° field of view

Locate α Cygni. Find its apparent rising, transit, and setting times.

Rising time: _____:_____ Transit time: _____:_____ Setting time: _____:_____

R.A. (2010): _____ Declination (2010): _____

Altitude: _____ Azimuth: _____

Note the differences between the times in Part 3 and Part 1, and record them in the following spaces:

Differences in rising times: _____ Differences in transit times: _____

Differences in setting times: _____

Were there any differences in the altitude and azimuth? _____. What were they?

Difference in altitudes: _____ Difference in azimuths: _____

Why do the results of Part 3 and Part 1 differ? _____
_____

4. Close the file and do not save, or continue with Exercise F.

Exercise F: Observing Longitudinal Differences in Time

1. Reset *TheSky* as follows:

    Location: Ball State Observatory
    Elevation: 322 meters
    Date: August 25, 2002
    Time: 21:00   Time Zone: +5.00
    Daylight savings adjustment option: "Not Observed"
    Orientation: Move To: R. A.: 18:56:32.1   Declination: 40° 11′ 8.7″ North
    Current Epoch   Orientation: Zoom to: 50° field of view

    Turn the Equatorial Grid off, and turn on the Horizon Grid.
    Make sure that the Bayer letters are displayed in the sky window.

    Locate α Lyrae. Find its apparent rising, transit, and setting times.

    Proper name: _____

    Rising time: _____:_____ Transit time: _____:_____ Setting time: _____:_____

    R.A. (2002): _____ Declination (2002): _____

    Altitude: _____   Azimuth: _____

2. Change the setup as follows:

    Location: Memphis, TN
    Elevation: 100 meters
    Date: August 25, 2002
    Time: 21:00   Time Zone = +5.00
    Daylight savings adjustment option: "Not Observed"
    Orientation: Move To: R. A.: 18:56:32.1   Declination: 40° 11′ 8.7″ North
    Current Epoch   Orientation: Zoom To: 50° field of view

    Locate α Lyrae. Find its apparent rising, transit, and setting times.

    Rising time: _____:_____ Transit time: _____:_____ Setting time: _____:_____

    R.A. (2002): _____ Declination (2002): _____

    Altitude: _____   Azimuth: _____

    Note the differences between the results of Part 2 and Part 1, and record them in the following spaces:

    Differences in rising times: _____   Differences in transit times: _____

    Differences in setting times: _____

    Were there any differences in the altitude and azimuth? _____. What were they?

    Difference in altitudes: _____   Difference in azimuths: _____

    Why do the results between Part 2 and Part 1 differ? _____
    _____

3. Change the setup as follows:

   Longitude: 90° 00′ 00″
   Elevation: 100 meters
   Latitude: 40° 11′ 59″ (Same as Ball State Observatory)
   Date: August 25, 2002
   Time: 21:00   Time Zone: +5
   Daylight savings adjustment option: "Not Observed"
   Orientation: Move To: R. A.: 18:56:32.1   Declination: 40° 11′ 8.7″ North
   Current Epoch   Orientation: Zoom To: 50° field of view

   Locate α Lyrae. Find its apparent rising, transit, and setting times.

   Rising time: _____:_____   Transit time: _____:_____   Setting time: _____:_____

   R.A. (2002): _____   Declination (2002): _____

   Altitude: _____   Azimuth: _____

   Note the differences between the results of Part 3 and Part 1, and record them in the
   following spaces:

   Differences in rising times: _____   Differences in transit times: _____

   Differences in setting times: _____

   Were there any differences in the altitude and azimuth? _____. What were they?

   Difference in altitudes: _____   Difference in azimuths: _____

   Why do the results between Part 3 and Part 1 differ? _____
   _____

4. Close the file and do not save, or continue with **Exercise G**.

Exercise G: Observing Longitudinal Differences in Time

1. Reset *TheSky* as follows:

   Location: Ball State Observatory
   Elevation: 322 meters
   Date: April 20, 2010
   Time: 05:20   Time Zone: +5.00
   Daylight savings adjustment option: "Not Observed"
   Orientation: Move To: R. A.: 19:55:30.5   Declination: 8° 55′ 15.0″ North
   Current Epoch   Orientation: Zoom To: 50° field of view

   Turn the Equatorial Grid off, and turn on the Horizon Grid.
   Make sure that the Bayer letters are displayed in the sky window.

   Locate α Aquilae. Find its apparent rising, transit, and setting times.

   Proper name: _____

   Rising time: _____:_____   Transit time: _____:_____   Setting time: _____:_____

R.A. (2010): _____ Declination (2010): _____

Altitude: _____ Azimuth: _____

2. Change the setup as follows:

   Location: Memphis, TN
   Elevation: 100 meters
   Date: April 20, 2010
   Time: 05:20   Time Zone = +5.00
   Daylight savings adjustment option: "Not Observed"
   Orientation: Move To: R. A.: 19:55:30.5   Declination: 8° 55′ 15.0″ North
   Current Epoch   Orientation: Zoom To: 50° field of view

   Locate α Aquilae. Find its apparent rising, transit, and setting times.

   Rising time: _____:_____ Transit time: _____:_____ Setting time: _____:_____

   R.A. (2010): _____ Declination (2010): _____

   Altitude: _____ Azimuth: _____

   Note the differences between the results of Part 2 and Part 1, and record them in the
   following spaces:

   Differences in rising times: _____   Differences in transit times: _____

   Differences in setting times: _____

   Were there any differences in the altitude and azimuth? _____. What were they?

   Difference in altitudes: _____   Difference in azimuths: _____

   Why do the results between Part 2 and Part 1 differ? _____
   _____

3. Change the setup as follows:

   Longitude: 90° 00′ 00″
   Elevation: 100 meters
   Latitude: 40° 11′ 59″ (Same as Ball State Observatory)
   Date: April 20, 2010
   Time: 05:20   Time Zone: +5
   Daylight savings adjustment option: "Not Observed"
   Orientation: Move To: R. A.: 19:55:30.5   Declination: 8° 55′ 15.0″ North
   Current Epoch   Orientation: Zoom To: 50° field of view

   Locate α Aquilae. Find its apparent rising, transit, and setting times.

   Rising time: _____:_____ Transit time: _____:_____ Setting time: _____:_____

   R.A. (2010): _____ Declination (2010): _____

   Altitude: _____ Azimuth: _____

Note the differences between the results of Part 3 and Part 1, and record them in the following spaces:

Differences in rising times: _____   Differences in transit times: _____

Differences in setting times: _____

Were there any differences in the altitude and azimuth? _____. What were they?

Difference in altitudes: _____   Difference in azimuths: _____

Why do the results between Part 3 and Part 1 differ? _____
_____

4.   Close the file and do not save, or continue with Exercise H.

Exercise H: Observing Longitudinal Differences in Time

1.   Reset *TheSky* as follows:

Location: Ball State Observatory
Elevation: 322 meters
Date: December 31, 2047
Time: 21:30   Time Zone: +5.00
Daylight savings adjustment option: "Not Observed"
Orientation: Move To: R. A.: 06:01:00.0   Declination: 10° 00′ 15.0″ North
Current Epoch   Orientation: Zoom To: 50° field of view

Turn the Equatorial Grid off, and turn on the Horizon Grid.
Make sure that the Bayer letters are displayed in the sky window.

Locate α Orionis. Find its apparent transit time.

Proper name: _____

Rising time: _____:_____ Transit time: _____:_____ Setting time: _____:_____

R.A. (2047): _____ Declination (2047): _____

Altitude: _____   Azimuth: _____

2.   Change the setup as follows:

Location: Memphis, TN
Elevation: 100 meters
Date: December 31, 2047
Time: 21:30   Time Zone = +6.00
Daylight savings adjustment option: "Not Observed"
Orientation: Move To: R. A.: 06:01:00.0   Declination: 10° 00′ 15.0″ North
Current Epoch   Orientation: Zoom To: 50° field of view

Locate α Orionis. Find its apparent transit time.

Rising time: _____:_____ Transit time: _____:_____ Setting time: _____:_____

R.A. (2047): _____ Declination (2047): _____

Altitude: _____ Azimuth: _____

Note the differences between the results of Part 2 and Part 1 and record them in the following spaces:

Differences in rising times: _____ Differences in transit times: _____

Differences in setting times: _____

Were there any differences in the altitude and azimuth? _____. What were they?

Difference in altitudes: _____ Difference in azimuths: _____

Why do the results between Part 2 and Part 1 differ? _____
_____

3. Change the setup as follows:

Longitude: 90° 00′ 00″
Elevation: 100 meters
Latitude: 40° 11′ 59″ (Same as Ball State Observatory)
Date: December 31, 2047
Time: 21:30   Time Zone: +5
Daylight savings adjustment option: "Not Observed"
Orientation: Move To: R. A.: 06:01:00.0   Declination: 10° 00′ 15.0″ North
Current Epoch   Orientation: Zoom To: 50° field of view

Locate α Orionis. Find its apparent transit time.

Rising time: _____:_____ Transit time: _____:_____ Setting time: _____:_____

R.A. (2047): _____ Declination (2047): _____

Altitude: _____ Azimuth: _____

Note the differences between the results of Part 3 and Part 1 and record them in the following spaces:

Differences in rising times: _____ Differences in transit times: _____

Differences in setting times: _____

Were there any differences in the altitude and azimuth? _____. What were they?

Difference in altitudes: _____ Difference in azimuths: _____

Why do the results between Part 3 and Part 1 differ? _____
_____

4. Close the file and do not save it.

If you answered that the results differed between Part 1 and Part 2, and Part 1 and Part 3, in the preceding exercises because of the different longitudes of the observers, then you are correct!

Determining the Sidereal Time

Exercises **I**, **J**, **K**, **L**, and **M** are designed to help you determine the sidereal time. You can determine the sidereal time for any location, any time of day, and for day of the year. Settings for time zone setting, elevation, and Daylight Savings Time (except **Exercise M**) remain the same as in the previous examples. In actuality, these settings would change to reflect a particular location on Earth. The settings for these exercises will not be changed, to only demonstrate the longitudinal effects.

Exercise I: Determining Sidereal Time for Different Locations on Earth

1. Start *TheSky* and set as follows:

   Location: Ball State Observatory
   Elevation: 322 meters
   Date: February 2, 2002
   Time: 21:00   Time Zone: +5.00
   Daylight savings adjustment option: "Not Observed"
   Orientation: Move To: R.A.: 05:15:00.0   Declination: 49° 30′ 0.0″ North
   Current Epoch

   What is the name of the red dotted line that passes through the field of view?

   _____ _____ _____

   Find the coordinate readout located near the bottom and just right of center in your sky window. It displays right ascension and declination of the cursor's position in the window. If you put the cursor on the vertical red dotted line (local celestial meridian for an observer at Ball State Observatory), then you will display the coordinates of any object located on the local celestial meridian. The sidereal time *is*, by definition, the right ascension of any object on the local celestial meridian. Put your cursor on the red dotted line! What is the sidereal time at the Ball State Observatory?

   Sidereal Time = ____ : ____ : ____

   There is a way to estimate the sidereal time at this location. Are there any bright stars located close to the local celestial meridian? If so does it have a proper name?

   Proper name: _____

   What are its equatorial coordinates?

   R.A. (2002): ____ᴴ ____ᴹ ____ˢ   Declination (2002): __ ____° ____′ ____″

   Could they help you determine the sidereal time? ____
   If so, how? _____

2. Reset *TheSky* to the following location:

   Location: Philadelphia, PA
   Elevation: 100 meters
   Date: February 2, 2002
   Time: 21:00   Time Zone: +5.00
   Daylight savings adjustment option: "Not Observed"
   Orientation: Move To: R.A.: 05:15:00.0   Declination: 49° 30′ 0.0″ North

Current Epoch

Put your cursor on the red dotted line again. Use the coordinate readout at the bottom of the sky window to determine the sidereal time for an observer in Philadelphia, PA.

Sidereal Time = _____:_____:_____

Now compute the difference in the sidereal times between the two locations.

Difference in sidereal time (Part 2 minus Part 1) = _____:_____

Why are the two times different? _____

3. Reset *TheSky* to the following location:

Location: 75th Longitude
Elevation: 100 meters
Date: February 2, 2002
Time: 21:00   Time Zone: +5.00
Daylight savings adjustment option: "Not Observed"
Orientation: Move To: R.A.: 05:15:00.0   Declination: 49° 30′ 0.0″ North
Current Epoch

Put your cursor on the local celestial meridian once again and determine the sidereal time for an observer located at the 75th longitude circle.

Sidereal Time = _____:_____:_____

Now compute the difference in time between the locations.

Difference in Sidereal Time (Part 3 minus Part 1) = ____:____

Difference in Sidereal Time (Part 3 minus Part 2) = ____:____

Why are the two time differences different? _____

4. Close the file and do not save, or continue with **Exercise J**.

Exercise J: Determining Sidereal Time for Your Location on Earth

1. Reset *TheSky* as follows:

Location: Ball State Observatory
Elevation: 322 meters
Date: March 15, 2005
Time: 20:45   Time Zone: +5.00
Daylight savings adjustment option: "Not Observed"
Orientation: Move To: R. A.: 07:45:38.7   Declination: 28° 00′ 57.8″ North
Current Epoch

What is the name of the red dotted line that passes through the field of view?

_____ _____ _____

Put your cursor on the red dotted line again. Use the coordinate readout at the bottom of the sky window to determine the sidereal time for an observer located at the Ball State Observatory.

Sidereal Time = _____:_____:_____

Are there any bright stars located close to the local celestial meridian near the zenith? If so, does it have a proper name? _____

Proper Name: _____

What are its equatorial coordinates?

R.A. (2005): ____^{H} ___^{M} ____^{S} Declination (2005): __ ____$^\circ$ ____' ____"

2. Reset *TheSky* to the following location:

Location: Memphis, TN
Elevation: 100 meters
Date: March 15, 2005
Time: 20:45   Time Zone: +5.00
Daylight savings adjustment option: "Not Observed"
Orientation: Move To: R. A.: 07:45:38.7   Declination: 28° 00' 57.8" North
Current Epoch

Put your cursor on the red dotted line again. Use the coordinate readout at the bottom of the sky window to determine the sidereal time for an observer located in Memphis, TN.

Sidereal Time = _____:_____:_____

Now compute the difference in time between the two locations.

Difference in Sidereal Time (Part 2 minus Part 1) = ____: ____

Why are the two times different? _____

3. Reset *TheSky* to the following location:

Location: 90th Longitude
Elevation: 100 meters
Date: March 15, 2005
Time: 20:45   Time Zone: +5.00
Daylight savings adjustment option: "Not Observed"
Orientation: Move To: R. A.: 07:45:38.7   Declination: 28° 00' 57.8" North
Current Epoch

Put your cursor on the red dotted line again. Use the coordinate readout at the bottom of the sky window to determine the sidereal time for an observer located on the 90th longitude circle.

Sidereal Time = _____:_____:_____

Now compute the difference in time between the two locations.

Difference in Sidereal Time (Part 3 minus Part 1) = ____:____

Difference in Sidereal Time (Part 3 minus Part 2) = ____:____

Why are the two time differences different? _____

4. Close the file and do not save, or continue with Exercise K.

Exercise K: Determining Sidereal Time for Your Location on Earth

1. Reset *TheSky* as follows:

Location: Ball State Observatory
Elevation: 322 meters
Date: July 4, 2025
Time: 21:30   Time Zone: +5.00
Daylight savings adjustment option: "Not Observed"
Orientation: Move To: R. A.: 14:35:45.0   Declination: 19° 30′ 5.0″ North
Current Epoch

What is the name of the red dotted line that passes through this field of view?

_____ _____ _____

Put your cursor on the red dotted line again. Use the coordinate readout at the bottom of the sky window to determine the sidereal time for an observer located at Ball State Observatory.

Sidereal Time = _____:_____:_____

Are there any bright stars located close to the local celestial meridian near the zenith? If so, does it have a proper name? _____

Proper Name: _____

What are its equatorial coordinates?

R.A. (2025): ____^{H} ___^{M} ____^{S} Declination (2025): __ ____° ____′ ____″

2. Now reset *TheSky* to the following location:

Location: Memphis, TN
Elevation: 100 meters
Date: July 4, 2025
Time: 21:30   Time Zone: +5.00
Daylight savings adjustment option: "Not Observed"
Orientation: Move To: R. A.: 14:35:45.0   Declination: 19° 30′ 5.0″ North
Current Epoch

Put your cursor on the red dotted line again. Use the coordinate readout at the bottom of the sky window to determine the sidereal time for an observer located in Memphis, TN.

Sidereal Time = _____:_____:_____

Now compute the difference in time between the two locations.

Difference in Sidereal Time (Part 2 minus Part 1) = ____ : ____

Why are the two times different? _____

3. Reset *TheSky* to the following location:

Set Location: 90th Longitude
Elevation: 100 meters
Date: July 4, 2025
Time: 21:30   Time Zone: +5.00
Daylight savings adjustment option: "Not Observed"
Orientation: Move To: R. A.: 14:35:45.0   Declination: 19° 30′ 5.0″ North
Current Epoch

Put your cursor on the red dotted line again. Use the coordinate readout at the bottom of the sky window to determine the sidereal time for an observer located on the 90th longitude circle.

Sidereal Time = _____ : _____ : _____

Now compute the difference in time between the two locations.

Difference in Sidereal Time (Part 3 minus Part 1) = ____ : __ __

Difference in Sidereal Time (Part 3 minus Part 2) = ____ : ____

Why are the two time differences different? _____

4. Close the file and do not save, or continue with Exercise L.

Exercise L: Determining Sidereal Time for Your Location on Earth

1. Reset *TheSky* as follows:

Location: Ball State Observatory
Elevation: 322 meters
Date: April 20, 2010
Time: 05:20   Time Zone: +5.00
Daylight savings adjustment option: "Not Observed"
Orientation: Move To: R. A.: 18:56:32.1   Declination: 40° 11′ 8.7″ North
Current Epoch

What is the name of the red dotted line that passes through this field of view?

_____ _____ _____

Put your cursor on the red dotted line again. Use the coordinate readout at the bottom of the sky window to determine the sidereal time for an observer located at Ball State Observatory.

Sidereal Time = _____:_____:_____

Are there any bright stars located close to the local celestial meridian near the zenith? If so, does it have a proper name? _____

Proper Name: _____

What are its equatorial coordinates?

R.A. (2010): ____^{H} ___^{M} ____^{S} Declination (2010): __ ____$^\circ$ ____$'$ ____$''$

2.   Reset *TheSky* to the following location:

Location: Philadelphia, PA
Elevation: 100 meters
Date: April 20, 2010
Time: 05:20   Time Zone: +5.00
Daylight savings adjustment option: "Not Observed"
Orientation: Move To: R. A.: 18:56:32.1   Declination: 40° 11′ 8.7″ North
Current Epoch

Put your cursor on the red dotted line again. Use the coordinate readout at the bottom of the sky window to determine the sidereal time for an observer located in Philadelphia, PA.

Sidereal Time = _____:_____:_____

Now compute the difference in time between the two locations.

Difference in Sidereal Time (Part 2 minus Part 1) = ____: ____

Why are the two times different? _____

3.   Now reset *TheSky* to the following location.

Location: 75th Longitude
Elevation: 100 meters
Date: April 20, 2010
Time: 05:20   Time Zone: +5.00
Daylight savings adjustment option: "Not Observed"
Orientation: Move To: R. A.: 18:56:32.1   Declination: 40° 11′ 8.7″ North
Current Epoch

Put your cursor on the red dotted line again. Use the coordinate readout at the bottom of the sky window to determine the sidereal time for an observer located on the 75th longitude circle.

Sidereal Time = _____:_____:_____

Now compute the difference in time between the two locations.

Difference in Sidereal Time (Part 3 minus Part 1) = ____:____

Difference in Sidereal Time (Part 3 minus Part 2) = ____:____

Why are the two time differences different? _____

4. Close the file and do not save , or continue with Exercise M.

Exercise M: Determining Sidereal Time for Your Location on Earth

1. Reset *TheSky* as follows:

   Location: Vandenberg Air Force Base, CA
   Longitude: 120° 31′ 05″ W   Latitude: 34° 44′ 35″ N
   Elevation: 100 meters
   Date: December 8, 2013
   Time: 11:30   Time Zone: +8.00
   Daylight savings adjustment option: "North America"
   Orientation: Move To: R.A.: 17:30:30.5   Declination: 35° 00′ 0.0″ North
   Current Epoch

   What is the name of the red dotted line that passes through this field of view?

   _____   _____   _____

   Put your cursor on the red dotted line again. Use the coordinate readout at the bottom of the sky window to determine the sidereal time for an observer located at the Vandenberg Air Force Base, CA.

   Sidereal Time = _____:_____:_____

   Are there any bright stars located close to the local celestial meridian near the zenith? If so, does it have a proper name? _____

   Proper Name: _____

   What are its equatorial coordinates?

   R.A. (2013): ____^{H} ___^{M} ____^{S} Declination (2013): __ ____° ____′ ____″

2. Reset *TheSky* to the following location:

   Location: Los Angles, CA
   Elevation: 100 meters
   Date: December 8, 2013
   Time: 11:30   Time Zone: +8.00
   Daylight savings adjustment option: "North America"
   Orientation: Move To: R. A.: 17:30:30.5 Declination: 35° 00′ 0.0″ North
   Current Epoch

Put your cursor on the red dotted line again. Use the coordinate readout at the bottom of the sky window to determine the sidereal time for an observer located in Los Angeles, CA.

Sidereal Time = _____:_____:_____

Now compute the difference in time between the two locations.

Difference in Sidereal Time (Part 2 minus Part 1) = ____:____

Why are the two times different? _____

3. Reset *TheSky* to the following location:

Location: 120th Longitude
Elevation: 100 meters
Date: December 8, 2013
Time: 11:30   Time Zone: +8.00
Daylight savings adjustment option: "North America"
Orientation: Move To: R. A.: 17:30:30.5   Declination: 35° 00′ 0.0″ North
Current Epoch

Put your cursor on the red dotted line again. Use the coordinate readout at the bottom of the sky window to determine the sidereal time for an observer located on the 120th longitude circle.

Sidereal Time = _____:_____:_____

Now compute the difference in time between the two locations.

Difference in Sidereal Time (Part 3 minus Part 1) = ____:____

Difference in Sidereal Time (Part 3 minus Part 2) = ____:____

Why are the two time differences different? _____

4. Close the file and do not save it.

If your answer to why the results between Part 1 and Part 2, and Part 1 and Part 3, in the previous exercises were because of the different longitudes of the observers, then you are right again!

It should have become apparent by now that if you know the right ascension of objects on the local celestial meridian, you know the sidereal time! As it turns out, it doesn't matter where you are located on Earth, the day of the year, or even the year; you can determine the sidereal time from your location. Using *TheSky* to ascertain the sidereal time at your location is a simple task. The previous exercises were designed to teach this and may have seemed repetitious.

If you have a star chart, then you can look for a star or some constellation on or near your local celestial meridian. This is a quick way to estimate the sidereal time; *TheSky* is the best way!

Knowing sidereal time is crucial in determining what constellation, star, planet, or any other celestial object might be visible for a given date at a specific location on Earth. *TheSky* makes determining sidereal time for any date and time simple and easy. Planning observing sessions also becomes easier. If an object's right ascension is within $\pm6^H$ of the sidereal time, then you know that the object is above your horizon. Knowing the sidereal time before observing makes the experience more productive and pleasant.

Determining Solar Time

Time differences seen in the transits of astronomical objects are not limited to the objects seen just in the night sky. You can also use *TheSky* to observe transits of the most obvious astronomical objects in the sky. Not only do we observe time differences in the transits of stars at different locations on Earth, but we also observe that the Sun itself doesn't transit the local celestial meridian at exactly noon each day. In fact, it does so only about four times during the year.

There are time differences in solar transits because Earth's orbit is not circular, but elliptical. As Earth revolves around the Sun, its distance from the Sun constantly changes. The Sun in turn exerts a net force on Earth that constantly changes. Earth responds to this net force exerted on it by the Sun by accelerating. That is, Earth speeds up and slows down as it orbits the Sun.

Earth moves faster when it is closer to the Sun and slower when it is farther away. It *moves fastest* in its orbit when it is *closest* to the Sun (*perihelion*), which occurs on or about January 3rd each year. It moves *slowest* in its orbit when it is *farthest away* from the Sun (*aphelion*), which occurs on or about July 3rd each year.

The motion and tilt of Earth affect our observations of the Sun and how we determine solar time. As the Sun appears to move eastward along the ecliptic, it reflects the motion of Earth around it. That is, the Sun's motion in the sky appears to speed up during January and slow down during July each year.

The 1° eastward motion discussed in Chapter 5 is an average value. During January each year the Sun appears to move a bit more than 1° per day, whereas during July each year the Sun appears to move a bit less than 1° per day along the ecliptic. The tilt of Earth's axis makes the Sun appear to move north and south of the celestial equator. When all this is factored into determining the solar time, it turns out that the rotating Earth is a poor timekeeper!

Astronomers and timekeepers avoid this problem by inventing a fictitious Sun called a mean Sun. This mean Sun moves along the celestial equator at a uniform rate rather than along the ecliptic. The rate of movement, at which the fictitious Sun moves along the celestial equator, is an average value of the apparent Sun's motion during the course of a year along the ecliptic. The mean Sun eliminates any irregularities observed in the apparent Sun's motion caused by the motion of Earth. The time interval between two successive passages of the mean Sun across a celestial meridian is defined as a *mean solar day*.

One manifestation of this irregular motion of Earth is the time at which the Sun appears to transit your local celestial meridian. Exercises N, O, P, and Q are designed to determine differences in transit times of the Sun across an observer's local celestial meridian. For doing these exercises it is imperative to set the date and times as indicated.

Exercise N: Determining Solar Time for Your Location on Earth

1. Start *TheSky* and set as follows:

    Location: Ball State Observatory
    Date: March 15, 2005
    Time: 12:00:00   Time Zone: +5.00
    Daylight savings adjustment option: "Not Observed"

    Find the Sun and center it in the sky window.

    What does the solid teal line through the Sun represent? _____

    What time does the Sun appear to transit? _____:_____

    Why isn't it 12:00 Noon? _____

2. Reset *TheSky* to the following location:

   Location: Philadelphia, PA
   Date: March 15, 2005
   Time: 12:00:00   Time Zone: +5.00
   Daylight savings adjustment option: "Not Observed"

   Find the Sun and center it in the sky window.

   What time does the Sun appear to transit? _____:_____

   Why isn't it 12:00 Noon again? _____

   What is the difference in the transit time (12:00 − observed transit time)? _____:_____

   Is the difference in time close to the equation of time value for this date in Table 7-1? _____

3. Close the file and do not save, or continue with Exercise O.

Exercise O: Determining Solar Time for Your Location on Earth

1. Reset *TheSky* as follows:

   Location: Ball State Observatory
   Date: June 1, 2010
   Time: 12:00:00   Time Zone: +5.00
   Daylight savings adjustment option: "Not Observed"

   Find the Sun and center it in the sky window.

   What does the solid teal line through the Sun represent? _____

   What time does the Sun appear to transit? _____:_____

   Why isn't it 12:00 Noon? _____

2. Reset *TheSky* to the following location:

   Location: Philadelphia, PA
   Date: June 1, 2010
   Time: 12:00:00   Time Zone: +5.00
   Daylight savings adjustment option: "Not Observed"

   Find the Sun and center it in the sky window.

   What time does the Sun appear to transit? _____:_____

   Why isn't it 12:00 Noon again? _____

   What is the difference in the transit time (12:00 − observed transit time)? _____:_____

   Is the difference in time close to the equation of time value for this date in Table 7-1? _____

3. Close the file and do not save, or continue with Exercise P.

Exercise P: Determining Solar Time for Your Location on Earth

1. Reset *TheSky* to the following location:

   Location: Ball State Observatory
   Date: September 15, 2015
   Time: 12:00:00   Time Zone: +5.00
   Daylight savings adjustment option: "Not Observed"

   Find the Sun and center it in the sky window.

   What does the solid teal line through the Sun represent? _____

   What time does the Sun appear to transit? _____:_____

   Why isn't it 12:00 Noon? _____

2. Reset *TheSky* to the following location:

   Location: Memphis, TN
   Date: September 15, 2015
   Time: 12:00:00   Time Zone: +6.00
   Daylight savings adjustment option: "Not Observed"

   Find the Sun and center it in the sky window.

   What time does the Sun appear to transit? _____:_____

   Why isn't it 12:00 Noon again? _____

   What is the difference in the transit time (12:00 – observed transit time)? _____

   Is the difference in time close to the equation of time value for this date in Table 7-1? _____

3. Close the file and do not save, or continue with Exercise Q.

Exercise Q: Determining Solar Time for Your Location on Earth

1. Reset *TheSky* to the following location:

   Location: Ball State Observatory
   Date: December 1, 2020
   Time: 12:00:00   Time Zone: +5.00
   Daylight savings adjustment option: "Not Observed"

   Find the Sun and center it in the sky window.
   What does the solid teal line through the Sun represent? _____

   What time does the Sun appear to transit? _____:_____

Why isn't it 12:00 Noon? _____

2. Reset *TheSky* to the following location:

    Location: Memphis, TN
    Date: December 1, 2020
    Time: 12:00:00   Time Zone: +6.00
    Daylight savings adjustment option: "Not Observed"

    Find the Sun and center it in the sky window.

    What time does the Sun appear to transit? _____:_____

    Why isn't it 12:00 Noon again? _____

    What is the difference in the transit time (12:00 – observed transit time)? _____

    Is the difference in time close to the equation of time value for this date in Table 7-1? _____

3. Close the file and do not save it.

This difference in transit times of the Sun can be determined for any of the standard longitudes for any day of the year. As previously mentioned, the Earth is a lousy timekeeper. It speeds up and slows down as it orbits the Sun; its tilt makes the Sun appear to move north and south of the celestial equator. All these things combined make these differences in time, noted in Table 7-1, more apparent.

It should also be evident from doing the previous exercises that the state of Indiana is actually closer to the Central Standard Time (CST) zone longitude. Indiana, like a few other states, has elected to keep its clocks set on Eastern Standard Time (EST) instead. In fact, it is only one of two states, in the continental United States that remains on the same time (at least most of it) all year round! That is, it chooses to remain on Eastern Standard Time throughout the year.

Even though Indianapolis, the state capital, is on Eastern Standard Time all year long, the rest of the state is *never* on the same time at any given time during the year. Counties in the southern part of the State, along the Ohio River, are on Eastern Daylight Savings time during the summer months and on Eastern Standard Time during the winter months. Counties in the northwestern part of the state, near Chicago, are just the opposite. During the summer months they are on Central Daylight Savings Time (same as Indianapolis), and during the winter months these counties are one hour behind the rest of the state (CST).

Leaping ahead, falling behind, or falling ahead, leaping behind—confused? No wonder people have so much difficulty with time!

Another manifestation of the anomalies of Earth's motion is the position of the Sun in the sky at the time it appears to transit celestial meridians. During the year the mean Sun sometimes runs ahead of the apparent Sun and at other times during the year the mean Sun runs behind the apparent Sun.

During the course of a year this difference varies from −14.2 minutes to +16.4 minutes. The *difference* between *apparent solar time* and *mean solar time* is defined as the *equation of time*. These differences have been calculated for the year 2001 and are displayed in Table 7-1. High-precision formulas taken from the *2001 Astronomical Almanac* were used to determine these values. Calculations done for every day of the year appear at either of the two following Websites:

    web006.pavilion.net/users/aghelyar/sundat.htm
    freepages.pavilion.net/users/aghelyar/sundat.htm
These sites also have more information concerning how these values are determined.

Table 7-1

The Equation of Time

(mean solar time − apparent solar time)

Date	Minutes
January 1	−3.1
15	−9.0
February 1	−13.5
15	−14.2
March 1	−12.6
15	−9.2
April 1	−4.2
15	−0.3
May 1	+2.8
15	+3.7
June 1	+2.4
15	−0.2
July 1	−3.6
15	−5.8
August 1	−6.4
15	−0.7
September 1	−0.4
15	+4.4
October 1	+9.9
15	+14.0
November 1	+16.4
15	+15.6
December 1	+11.4
15	+5.4

If you graph the differences in time plotted against the declination of the Sun for several dates throughout the year, the resulting graph is known as an analemma. Figure 7-5 depicts *TheSky's* rendition of an analemma. In fact, with persistence you can photograph this phenomenon by taking multiple exposures on a single frame of photographic film.

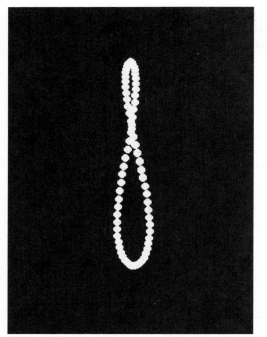

Figure 7-5 Analemma as shown in *TheSky*

To do so, you need a permanently mounted camera so that you can take a picture of the Sun's position in the sky at regular intervals throughout the year. The photograph must be taken at the same time of day. An excellent and detailed explanation of the analemma appears at the following Website: www.analemma.com.

In *TheSky* software the multimedia file named Analemma.sky illustrates an animated version of the analemma displayed in Figure 7-5. Exercise R displays the sky window with the Sun's location at the same time of day throughout an entire year for an observer in Denver, Colorado. The dates for this sky window are March 20, 1999, to March 19, 2000. The approximate time of day is 12:00 noon MST.

Exercise R: Observing the Analemma from Denver, Colorado

1. Open the file named Analemma.sky in the folder: C:\Program Files\Software Bisque\TheSky\User\Documents.

2. Click the ▶▶ button on the Time Skip toolbar.

3. To see the entire analemma, click the "Move Up" arrow (⬆) on the Orientation toolbar once.

4. Turn on the equatorial grid and horizon grid and watch the Sun's coordinates change throughout the year.

5. Close the file and do not save it, or leave it open and continue with Exercise S.

It is interesting to change your location to observe the analemma. The analemma is not something that happens at a particular latitude and longitude. It can be seen, for the most part, from every location on Earth. If you turn on the horizon grid while observing the analemma, it becomes apparent that the altitude of the Sun changes for any observer throughout the year. If you turn on the equatorial grid, you will notice that the right ascension and declination of the Sun also changes throughout the year.

Exercises S, T, U, V, and W let you view the analemma from different locations on Earth. Exercise S takes you to the north geographic pole while Exercise T takes you to the south geographic pole. Exercise U takes you to the geographic equator while Exercise V takes you to the Tropic of Cancer. Exercise W takes you to the Tropic of Capricorn.

Exercise S: Observing the Analemma from Earth's North Pole

1.   Open the multimedia file named Analemma.sky.

2.   Go to Data at the top of the sky window toolbar, then click Site Information, then click the Location tab and Open file "*Cities outside USA.loc*." Scroll down the indexed menu to the North Pole.

3.   Close the box and click the "South" button (�) to view the analemma.

4.   Click the ▶▶ button and observe.

5.   Turn on the equatorial grid and horizon grid and watch the Sun's coordinates change throughout the year.

6.   Close the file and do not save it, or leave it open and continue with Exercise T.

What you see from the North Pole is that portion of the analemma that is above the observer's horizon between mid-March and mid-September. The Sun's declination has positive values between 0° to +23½°. For about five and a half months, the North Pole is in total darkness throughout the year. For about two weeks or so it is in twilight and the rest of the time the Sun is above the horizon for the entire day (24^H).

Exercise T: Observing the Analemma from Earth's South Pole

1.   Open the multimedia file named Analemma.sky.

2.   Go to Data on the toolbar at the top of the sky window, then click Site Information, then click the Location tab and Open file "*Cities outside USA.loc*." Scroll down the indexed menu to the South Pole.

3.   Close the box and click the "North" button (⟨N⟩) to view the analemma.

4.   Click the ▶▶ button and watch.

5.   Turn on the equatorial grid and horizon grid and watch the Sun's coordinates change throughout the year.

6.   Close the file and do not save it, or leave it open and continue with *Exercise U*.

What you see at the South Pole is the portion of the analemma that is above an observer's horizon between mid-September and mid-March. The Sun's declination has negative values between 0° to −23½°. For about five and a half months, the South Pole is in total darkness throughout the year. For about two weeks or so it is in twilight and the rest of the time the Sun is above the horizon for the entire day (24^H).

Exercise U: Observing the Analemma from Earth's Equator

1. Open the multimedia file *Analemma.sky*.

2. Go to the Data at the top of the sky window, then click Site Information, then click the Location tab and Open file "*Cities outside USA.loc.*" Scroll down the indexed menu to the geographic equator.

3. Did you find the equator? If so, click it and continue as you did in the previous exercises. Otherwise, go to the Data, Site Information on the toolbar at the top of the sky window and set the latitude to that of the geographic equator, 0°. Longitude doesn't matter in this exercise!

4. Click Apply and Close the box.

5. Click the "Zenith" button (⊠) to view the analemma.

6. Click Orientation on the toolbar at the top of the sky window and Zoom To the Naked Eye. This resets the sky window to the same scale as the previous sky window.

7. Click the ⊠ button.

8. Turn on either the equatorial grid or horizon grid and watch the Sun's coordinates change throughout the year.

9. Close the file and do not save it, or leave it open and continue with Exercise V.

Because the geographic equator is not in *TheSky's* database you will have to add it to the database. For an observer at the equator the Sun is at the zenith twice a year, once at the vernal equinox and once at the autumnal equinox. At other times of the year, it is either north or south of the zenith point. The Sun appears to be 23° 26' north of the zenith point on the first day of summer and 23° 26' south of the zenith point on the first day of winter. Turn on the equatorial grid and use either the ⊠ button or the ⊠ button to verify this. Remember that the analemma is something that can be viewed over an entire year from anywhere on Earth.

Exercise V allows you to view the analemma from Havana, Cuba, or Syene, Egypt. These places are located near the Tropic of Cancer, 23 ½° north of the geographic equator.

Exercise V: Observing the Analemma from Near the Tropic of Cancer

1. Open the multimedia file named Analemma.sky.

2. Go to the Data at the top of the sky window, then click Site Information, then click the Location tab and Open file "*Cities outside USA.loc.*" Scroll down the indexed menu to Havana, Cuba.

3. Click Apply and Close the box.

4. Click the "Zenith" button (⊠) to view the analemma.

5. Click on Orientation at the top of the sky window then on Zoom To the Naked Eye.

6. Click the ⊠ button and watch.

7. Turn on the equatorial grid and horizon grid and watch the Sun's coordinates change throughout the year.

8. Close the file and do not save it, or leave it open and continue with Exercise W.

If an observer travels to the Tropic of Cancer (Syene, Egypt, for instance), the Sun's position in the sky is directly overhead at the zenith on the summer solstice at midday. The Sun is at its northernmost declination on the analemma on this date. For any observer on the Tropic of Cancer, the Sun is at the zenith only once during the year, at the summer solstice. Havana, Cuba is the only location in the database that is close to the latitude of Syene, Egypt or the Tropic of Cancer.

Exercise W: Observing the Analemma from near the Tropic of Capricorn

1. Open the multimedia file named Analemma.sky.

   Go to the Data at the top of the sky window, then click Site Information, then click the Location tab and Open file "*Cities outside USA.loc.*" Scroll down the indexed menu and search for the Tropic of Capricorn. You may have to set the latitude to that of the Tropic of Capricorn, 23½° south latitude. Longitude doesn't matter!

2. Click Apply and Close the box.

3. Click the Zenith button [z] to view the analemma.

4. Go to the Orientation at the top of the sky window and Zoom To Naked Eye to reset your sky window.

5. Click [▶▶] button and watch.

6. Turn on the equatorial grid or the horizon grid and watch the Sun's coordinates change throughout the year.

7. On what date would you observe the Sun at your zenith at this location? _____

8. Close the file and do not save it.

If an observer travels to the *Tropic of Capricorn*, the Sun's position in the sky is directly overhead at the zenith, on the winter solstice at midday. The Sun is at its southernmost declination on the analemma on this date. For any observer on the Tropic of Capricorn, the Sun is at the zenith only once during the year, at the winter solstice.

The main thing to remember while exploring this window or any other sky window is to be certain not to save this particular sky window when you decide to close it. If you do want to save this or any other sky window, remember to use the Save As option in the File Menu on the toolbar at the top of the sky window. The file then can be safely saved under a different filename.

Chapter 7

TheSky Review Exercises

Use *TheSky* to answer the following questions:

TheSky Exercise 1: Measuring Solar Time

Before starting *TheSky* make your settings to the following:

Date: January 15th   Time: 12:13
Daylight savings adjustment <u>o</u>ption: "Not Observed"
Location: 75° 00′ 00″ W   40° 00′ 00″ N
Time <u>Z</u>one: +5.00
Ele<u>v</u>ation: 100 meters

Turn off the daytime sky if it is on.

Find the Sun and center it.

Notice that the Sun is aligned with the star α Aquilae which is about 30° N of the Sun.
Set the Time Skip increment to $23^H 56^M 04^S$ (Sidereal Day).

Use the "Step Forward" button to advance the sky forward in time by 1 sidereal day.

1. How much does the Sun appear to move in the sky on this date? _____° _____′ _____″

2. What direction does the Sun appear to move with respect to the background stars? _____

3. Which object appears to cross the local celestial meridian at the 75th longitude circle on January 16th, the Sun or Altair? _____

4. How much earlier does this object cross the local celestial meridian each day? _____

5. What time does the Sun transit the local celestial meridian at this location? _____:_____

6. Is the difference in transit time in Question 5 (12:00 − observed transit time) consistent with the value in Table 7-1? _____

TheSky Exercise 2: Measuring Solar Time

Now make the following changes:

Date: January 15th   Time: 12:13
Daylight savings adjustment <u>o</u>ption: "Not Observed"
Location: 90° 00′ 00″ W   40° 00′ 00″ N
Time <u>Z</u>one: +6.00
Ele<u>v</u>ation: 100 meters

Find the Sun and center it.

1. Is the Sun aligned with the star α Aquilae from this location? _____

2. How much does the Sun appear to move in the sky on this date from this location?
   _____ ° _____ ' _____ "

3. What direction does the Sun appear to move with respect to the background stars from this location? _____

4. Which object appears to cross the local celestial meridian at this location on January 16th, the Sun or Altair? _____

5. How much earlier does this object cross the local celestial meridian each day? _____

6. What time does the Sun transit the local celestial meridian at this location? _____:_____

7. Is the difference in transit time in Question 6 (12:00 – observed transit time) consistent with the value in Table 7-1? _____

Chapter 7

TheSky Review Questions

1. The great circle passing through the south point on the horizon, the zenith, and the north point on the horizon is called the _____ _____ _____

2. What object in the sky do astronomer use to define apparent solar time? _____

3. Which point in the sky do astronomers use to define sidereal time? _____

4. An hour circle on the celestial sphere is exactly like a(n) _____ on the surface of the Earth.

5. If two locations on the surface of the Earth have the same local times, they can be said to have the same _____.

6. If the Sun and a star cross your local celestial meridian at exactly noon today, which object will appear to cross your local celestial meridian *first* at exactly noon tomorrow? _____

7. How much earlier will the transit take place tomorrow in Question 6? _____

8. Do the constellations (stars) appear to shift east or west each day as the seasons go by? _____

9. How much do the stars in Question 8 appear to shift each day? _____

10. Why do the stars in Question 8 appear to shift? _____

11. Because the Earth is curved, would an observer on the eastern edge of the Eastern Standard Time zone see an object appear to rise before or after an observer at the 75th longitude circle? _____

12. Similarly, does an observer on the western edge of the Eastern Standard Time zone see an object appear to rise before or after an observer at the 75th longitude circle? _____

    Although objects appear to rise or set at approximately the same local solar time for each observer, their watch times will differ for each observer at these different locations.

13. The hour angle of the apparent Sun is defined as _____ time.

14. The hour angle of the vernal equinox is defined as _____ time.

Chapter 8

Seasons in *TheSky*

The Seasons

We hope the complex topic of time seems clearer now. There are many things that you must know in order to make your observations productive. The beauty of using *TheSky* is that it does most of the hard work for you.

Now it's time to ask practical questions, like when does spring or autumn begin in Cleveland or when does the summer or winter solstice happen in Sydney this year? Usually, people associate seasons with the location of the Sun in the sky. You know, of course, that this is not true: the seasons are determined not by where the Sun is located in the sky but by where Earth is in its orbit. The Sun's position in the sky is one of several manifestations that occur due to Earth's physical location in space. The location of Earth in its orbit and its tilt *do*, in fact, determine where the Sun *is* on the ecliptic.

Many people associate the seasons with climatic changes on Earth. Our most reliable resource for finding the day of the year is the calendar, not the weather conditions. This is apparent for someone who lives in the northern United States in spring. Many cities in the Midwest enjoy springtime flowers and flowering trees, while at the same time those who live in the most northerly regions are subjected to cold and snowy conditions.

In Chapter 5, we discussed seasons as points in space. They were given celestial coordinates or positions of right ascension and declination on the celestial sphere. Now we will discuss them as points in time.

If you observed Earth from some distance away in space, you would easily see its motion around the Sun. As it travels around the Sun, its axis of rotation clearly remains oriented in the same direction in space. From the perspective of space, Earth as it travels in its orbit is not oriented upright. It appears "tilted" by an angle of 23° 26′.

The main cause of the seasons on Earth is this 23° 26′ angular difference between Earth's axis of rotation and its orbital plane. This tilt is known as the *obliquity of the ecliptic*. Astronomers measure this angle as the angle between Earth's equator and the plane of its orbit around the Sun. This angle is also the angle between the ecliptic north pole (point on the celestial sphere where Sun's axis of rotation points) and the north celestial pole (point on the celestial sphere where Earth's axis of rotation points). Earth's orientation in space and its orbital motion around the Sun are illustrated in Figure 8-1.

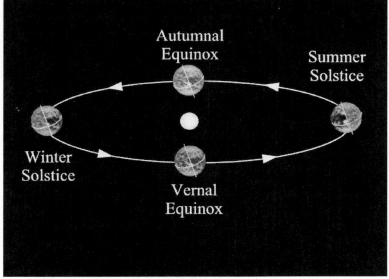

Figure 8-1 Obliquity of the ecliptic

As Earth revolves around the Sun, its axis of rotation remains rigid or fixed in space. That is, Earth's axis of rotation always points in the same direction in space, toward the north celestial pole. As a consequence of this, it points very close to the bright star Polaris, our North Star.

Earth's orientation in space does change, but over very long periods of time. The Earth spins like a top about its axis. And, like a top, it wobbles and changes its orientation.

The wobbling of Earth's axis is called nutation. If you look at an Earth globe, you will notice that the continental masses on the surface of Earth are not evenly distributed around the globe. As a result of this, Earth is unbalanced. Like an unbalanced tire on a car, it wobbles and shakes the car—in this case, the Earth. The wobbling of Earth's axis takes about 18.6 years to complete one cycle. The amount of wobble is only 9 arcseconds in angle projected onto the celestial sphere. As a result of Earth's nutation, the celestial sphere appears to wobble. This motion, of course, is so subtle that it is not observed except in large telescopes.

Another motion of Earth is the change in orientation of its axis. This motion is called precession. The fact that Earth rotates on its axis causes a bulge to occur in its oceans in the equatorial region. The Moon and Sun gravitationally tug on this bulge. Earth responds and would fall over if it were not spinning. Precession is simply a manifestation of Earth's stability in its orientation in space as it orbits the Sun. The time period for Earth to complete one change in orientation is 26,000 years.

These motions are two of Earth's lesser-known *true* motions in addition to rotation and revolution. Earth's precession has a very profound effect on the equatorial coordinates of right ascension and declination of objects in the sky and on the calendar. *TheSky* lets you move ahead or backward in time. In doing so, it is easy to see the effects of precession. By moving ahead in time several thousand years, it becomes obvious that Polaris will not always be our North Star.

Try to imagine, while standing on Earth, the motion produced in the sky while Earth moves around the Sun. Revolution is the motion that is responsible for the apparent motion of the Sun through the constellations of the zodiac. This apparent motion is eastward in the sky at a rate of about 1° per day.

The Sun appears to make one circuit of the sky along the ecliptic each year. The tilt of Earth causes the Sun to appear to move both north and south of the celestial equator during this period of time. It has taken humanity thousands of years to fully understand these motions. But, only in the last few hundred years have people fully have begun to understand the nature of these motions and how they are related to the motions in the sky and the seasons on Earth.

Vernal Equinox

If you recall, as we trace the ecliptic through the stars it eventually intersects the celestial equator at two points in space. These points are known as the equinoxes. One intersection or equinox is where the Sun moves from the Southern Hemisphere of the sky into the Northern Hemisphere. This event takes place on or about March 21 each year and is known as the vernal equinox.

If the Sun happens to be located at the vernal equinox, then this point in space actually becomes a point in time. That is, it *is* the first day of spring in the Northern Hemisphere. The equatorial coordinates of the Sun on the celestial sphere at the time of the vernal equinox are right ascension $00^H 00^M 00^S$ and declination of $00° 00' 00''$. In other words, the Sun's location on the celestial sphere is precisely on the celestial equator.

Figure 8-2 clearly shows the location of the vernal equinox in *TheSky*. The view is from Ball State Observatory with the vernal equinox positioned on the local celestial meridian. The altitude of this point is about 50° and its azimuth is 180°. At the bottom of the window, the right ascension coordinates are displayed.

If you toggle on the constellation boundaries, this point appears to be located in the western part of the constellation of Pisces. In about 500 or 600 years, this point will move into the constellation of Aquarius and will remain in this constellation for roughly 2000 years. This apparent motion is a result of Earth's precession.

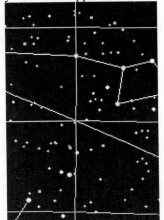

Figure 8-2 Location of the vernal equinox as shown in *TheSky*

Remember that the vernal equinox is the *zero point* for the coordinate of right ascension. All the equatorial coordinates of stars and deep sky objects are measured with respect to this point. If it moves, then the coordinates of objects in the sky change. *TheSky* calculates the new coordinates automatically. Only when a comparison is made between the coordinates in different epochs does it become apparent that Earth is changing its orientation in space. The sky window in Figure 8-2 may be viewed in the multimedia file named VernalEquinox.sky. It is interesting and fun to move ahead or back in time and to imagine how the sky appeared to our ancestors, or how it will appear in the future. It also lets us compare the coordinates of objects.

Exercises A, B, C, D, E, F, G, and H, which follow are designed to help you to determine when spring occurs in the Northern Hemisphere and how the equatorial coordinates of objects are affected by Earth's precession for a particular year.

Exercise A: Determining When the Vernal Equinox Will Occur in 2010 C.E.

1. Run *TheSky* and set as follows:

   Date: January 1, 2010
   Time: 00:15:00 A.M. EST

2. Click <u>A</u>pply, then Close.

3. Find the Sun.

4. Click on the General tab in Object Information window and scroll down the information window.

5. At the bottom of the information window, you will find seasonal information.

6. When will the vernal equinox occur in 2010 C.E.? _____
   (the answer is March 20th at 12:33 P.M. EST)

7. Close the file and do not save it, or continue with Exercise B.

Exercise B: Determining When the Vernal Equinox Occurred in 52 C.E.

1. Run *TheSky* and set as follows:

    Date: June 1, 52
    Time: 12:00 EST

2. Find the Sun.

3. When did the vernal equinox occur in 52 C.E.? _____
    (the answer is March 22nd at 4:04 A.M. EST)

4. Close the file and do no*t* save it, or continue with Exercise C.

Exercise C: Determining When the Vernal Equinox Occurred in 2735 B.C.E.

1. Run *TheSky* and set as follows:

    Date: October 4, −2734
    Time: 12:00 EST

2. Find the Sun.

3. When did the vernal equinox occur in 2735 B.C.E.? _____

4. Close the file and do not save it, or continue with Exercise D.

Exercise D: Determining When the Vernal Equinox Will Occur in 9999 C.E.

1. Run *TheSky* and set as follows:

    Date: September 15, 9999
    Time: 3:00 P.M. EST

2. Find the Sun.

3. When will the vernal equinox occur in 9999 C.E.? _____

4. Close the file and do not save it, or continue with Exercise E.

Exercise E: Determining the Equatorial Coordinates in 2010 C.E.

1. Run *TheSky* and set as follows:

    Date: January 10, 2010
    Time: 8:00 P.M. EST

2. Find Sirius and center it.

3. What are the right ascension and declination of Sirius?

R.A. = _____ ^H _____ ^M _____ ^S   Declination = __ _____ ° _____ ' _____ ''

4. Now find Polaris and center. Click on the [▸▸] button and observe the motion. Turn on the equatorial grid. Describe what you see. _____
_____

5. Close the file and do not save it, or continue with Exercise F.

Exercise F: Determining the Equatorial Coordinates in 52 C.E.

1. Run *TheSky* and set as follows:

   Date: January 10, 52
   Time: 8:00 P.M. EST

2. Find Sirius.

3. What are the right ascension and declination of Sirius?

   R.A. = _____ ^H _____ ^M _____ ^S   Declination = __ _____ ° _____ ' _____ ''

4. How do the coordinates compare with Exercises E? _____

5. Are they the same or different? _____

6. Now find Polaris and center it. Click on the [▸▸] button and observe the motion. Turn on the equatorial grid. Describe what you see. _____
_____

7. Close the file and do not save it, or continue with Exercise G.

Exercise G: Determining the Equatorial Coordinates in 9999 C.E.

1. Run *TheSky* and set as follows:

   Date: January 10, 9999
   Time: 8:00 P.M. EST

2. Find Sirius and center it.

3. What are the right ascension and declination of Sirius?

   R.A. = _____ ^H _____ ^M _____ ^S   Declination = __ _____ ° _____ ' _____ ''

4. How do the coordinates compare with those in Exercises E and F? _____
_____

5. Are they the same or different? _____

6. What do you notice about the location of Sirius with respect to your horizon? _____

7. Now find Polaris and center it. Click on the ⏩ button and observe the motion. Turn on the equatorial grid. Describe what you see. _____

8. Close the file and do not save it, or continue with Exercise H.

Exercise H: Determining the Equatorial Coordinates in 2735 B.C.E

1. Run *TheSky* and set as follows:

   Date: January 10, −2734
   Time: 8:00 P.M. EST

2. Find Sirius and center it.

3. What are the right ascension and declination of Sirius?

   R.A. = ____^H ____^M ____^S Declination = __ _____° _____′ _____″

4. How do the coordinates compare with those in Exercises E, F, and G? _____
   _____

5. Are they the same or different? _____

6. What do you notice about the location of Sirius with respect to your horizon? _____
   _____

7. Find Polaris and center it. Click on the ⏩ button and observe the motion. Turn on the equatorial grid. Describe what you see. _____
   _____

8. Is the motion of the sky about Polaris? _____

9. What star is located near the north celestial pole? _____
   Note: The Egyptians used this star as their North Star.

10. Close the file and do not save it.

The position of the Sun on the celestial sphere is directly over Earth's equator at the time of either equinox. The Sun's position in the sky on the vernal equinox in March is shown in Figure 8-3.

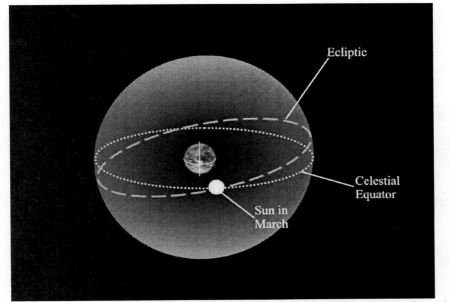

Figure 8-3 Position of Sun on the celestial sphere on the vernal equinox in March

If Earth is viewed from space at the time of the vernal equinox, its axis of rotation neither points toward or away from the Sun. In fact, the Sun shines directly on the Earth's equator and equally on both hemispheres, as shown in Figure 8-4. The length of days and nights are nearly equal at the time of the vernal equinox.

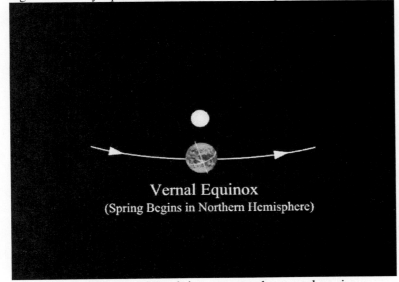

Figure 8-4 Position of Earth in space on the vernal equinox

Two factors contribute to seasons on Earth throughout the year. The first is the length of time the Sun spends above an observer's horizon. The second is the Sun's angular distance above the horizon, or its altitude. Both are a direct result of Earth's tilt.

From the time that the Sun appears to rise in the east each day, its altitude constantly changes in the sky. At midday, it reaches its highest altitude above the horizon. At this point, it is due south on the local celestial meridian. After midday, the Sun's altitude gets lower until it appears to set in the west. If you watch the apparent sunrise and sunset at the time of the vernal equinox, the Sun appears to rise *due east* and appears to set *due west* on the horizon.

Figure 8-5 illustrates the Sun's position on the eastern horizon on March 19, 2092, for an observer located at Ball State Observatory at 6:54 A.M. EST. The Sun reaches the vernal equinox point at precisely 9:35 A.M. EST on this date. Notice that the equatorial grid is displayed. The sky window plainly shows the yellow dotted line, representing the ecliptic, intersecting the celestial equator near the East Point on the horizon. If you travel to Earth's equator and observe the apparent sunrise there, the Sun appears to rise due east at that location.

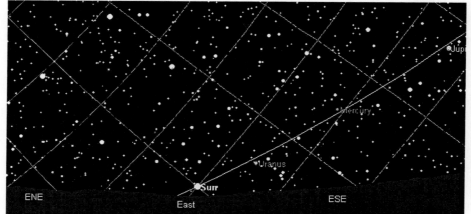

Figure 8-5 Apparent sunrise on March 19, 2092, from Ball State Observatory

Figure 8-6 illustrates the apparent sunrise for an observer located at the geographic equator.

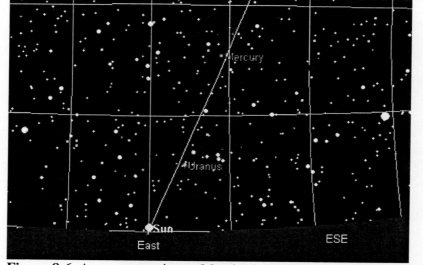

Figure 8-6 Apparent sunrise on March 19, 2092, at the geographic equator

The altitude of the Sun at midday on the vernal or autumnal equinox is about 50° for an observer located at the 40° north latitude circle. The altitude of the Sun at midday can be found for *any* observer in the Northern Hemisphere by applying the following relationship:

$$\text{Altitude}_{Sun} = 90° - \text{Latitude}_{observer}$$

The position of the Sun on March 19, 2092, at midday at Ball State Observatory is illustrated in Figure 8-7. Notice that the horizon grid is displayed. The Sun's position on the local celestial meridian is near the 50° altitude circle. For an observer located at Ball State Observatory at 12:48 P.M. EST, the actual altitude of the Sun is 49° 52′ 05″ as found in the Object Information window.

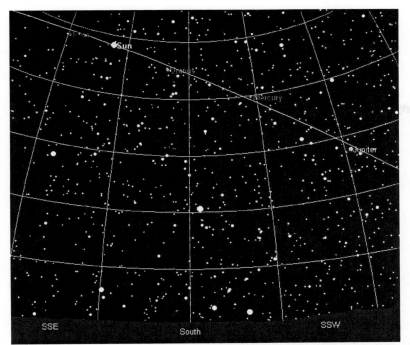

Figure 8-7 Altitude of Sun at midday, March 19, 2092, at Ball State Observatory

The position of the Sun for an observer at a different location on Earth on March 19, 2092, at midday, is illustrated in Figure 8-8. Notice that the horizon grid is once again displayed. It is quickly seen that the Sun's position on the local celestial meridian is near the 90° altitude circle. Where is the observer located?

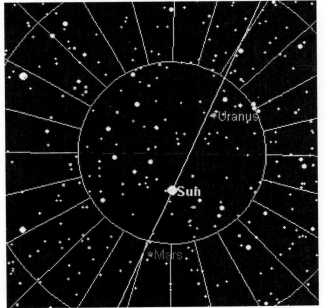

Figure 8-8 Altitude of Sun, midday, at a location different from that of Figure 8-7 on March 19, 2092

The answer, of course, is the geographic equator!

Summer Solstice

Each day that follows after the vernal equinox, the Sun moves farther into the Northern Hemisphere of the sky. A few days after the vernal equinox, the Sun's motion is both eastward and northward along the ecliptic. In the Northern Hemisphere, days become longer and warmer, and those in the Southern Hemisphere become shorter and cooler. It becomes apparent that the rising point of the Sun is moving farther north on the eastern horizon. To view this change of position in the Sun's rising point on the eastern horizon, complete Exercise I, which follows.

Exercise I: Observing Position of Sun's Rising Point on Eastern Horizon

1. Run *TheSky*.

2. Open the multimedia file named Season1.sky.

3. Click on the [▶▶] button. The time step is 2 days.

4. After 4 seconds, click the [■] button and observe the position of the yellow dotted line on the horizon. Described what happened. _____
    _____

5. Close the file.

The time sequence, in Exercise I, begins at the apparent sunrise on March 19, 2092, and continues through June 20, 2092. Notice that the Sun's altitude gets higher in the sky almost immediately. This is because the Sun appears to rise earlier each day from March until June. The ecliptic appears to move northward along the eastern horizon—watch the yellow dotted line.

Viewed from space during this time, the northern half of Earth's axis of rotation begins to align itself on the Sun. Earth has moved 90° in its orbit from the vernal equinox. The sunlight begins to shine more directly on the Northern Hemisphere, making the days longer and warmer.

The track of Earth's position during the time from the vernal equinox to the summer solstice can be seen in Figure 8-9. As Earth revolves around the Sun, it does *not* lean toward the Sun. Its orientation in space has remained this way all along.

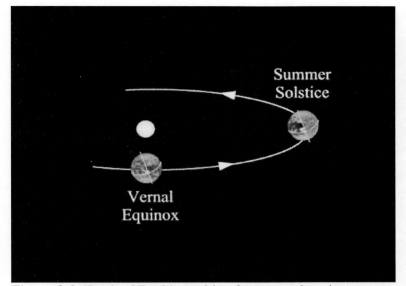

Figure 8-9 Track of Earth's position from vernal equinox to summer solstice

This position of Earth in space at the time of the summer solstice is illustrated in Figure 8-10. As Earth moves in its orbit from the vernal equinox to the location of the summer solstice, an interesting phenomenon appears on the eastern horizon. The apparent rising point of the Sun changes throughout this time interval—it begins to move farther northward. It reaches its most northerly position on the horizon on or about June 20th each year. At the most northerly point on the eastern horizon, the Sun appears to stop its northward motion. This point on the horizon where the Sun appears to stop is called the solstice point. The word *solstice* comes from a Greek word meaning, "Sun standing still" on the horizon. This is seen in the multimedia file named Season1.sky.

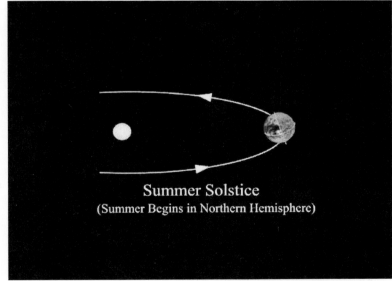

Figure 8-10 Position of Earth in space on the summer solstice

The apparent sunrise is illustrated in Figure 8-11 for an observer located at Ball State Observatory on June 20, 2092, at 5:30 A.M. EST. The Sun reaches the summer solstice point at precisely 2:18 A.M. EST on this date. At this instant, the Sun's equatorial coordinates on the celestial sphere are right ascension $6^H 00^M 00^S$ and declination $+23° 26'$.

Notice that the equatorial grid is displayed and the yellow dotted line, representing the ecliptic, is approximately 31° north of the East Point at this location.

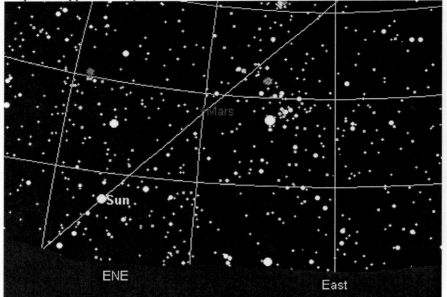

Figure 8-11 Apparent sunrise on June 20, 2092, at Ball State Observatory

If an observer is located at the geographic equator, however, the apparent rising point of the Sun is exactly 23° 26′ north of the East Point. The apparent sunrise for an observer located at the geographic equator is displayed in Figure 8-12.

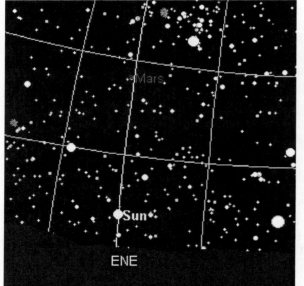

Figure 8-12 Apparent sunrise at geographic equator on June 20, 2092

Figure 8-13 illustrates the position of the Sun on the celestial sphere on the summer solstice in June. It is obvious that sunlight shines most directly on the Northern Hemisphere in this figure.

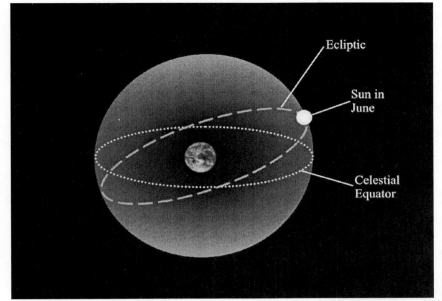

Figure 8-13 Position of Sun on the celestial sphere on the summer solstice in June

The Sun has the highest altitude at midday on the summer solstice for any observer located in the Northern Hemisphere. Figure 8-14 shows the Sun at midday for an observer located at Ball State Observatory on June 20, 2092. Notice that the horizon grid is displayed. The altitude of the Sun is about 73° on June 20th on this date. If you look in the Object Information window for this date, you can find the altitude of the Sun at midday. Is it about 73° at this location? The location of the summer solstice point may be viewed in the multimedia file named SummerSolstice.sky.

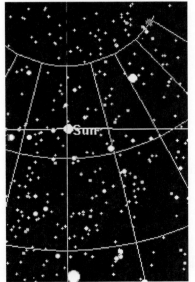

Figure 8-14 Altitude of Sun at Midday, June 20, 2092, at Ball State Observatory

In fact, for any observer who is located on the 23° 26′ north latitude circle (Tropic of Cancer), at midday, the Sun is located at their zenith. Exercise J illustrates the position of the Sun at midday on the Tropic of Cancer on June 20, 2092.

Exercise J: Observing the Sun at Midday on the Tropic of Cancer

1. Run *TheSky*.

2. Open the multimedia file named Season2.sky.

3. Set time: 12:43 P.M. EST.

4. Set latitude: 23° 26′ N.

5. Turn on the horizon grid.

6. Find the Sun and center.

7. Does your computer screen look like the one displayed in Figure 8-15?

8. What is the altitude of the Sun? _____°

9. Close the file and do not save it.

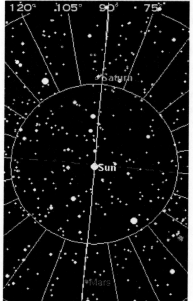

Figure 8-15 Sun's location in the sky at midday on the Tropic of Cancer

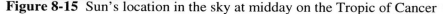

Exercises K, L, M, and N, which follow are designed to help you determine when summer occurs in the Northern Hemisphere for a particular year.

Exercise K: Determining When the Summer Solstice Will Occur in 2010 C.E.

1. Run *TheSky* and set as follows:

   Date: February 2, 2010
   Time: 8:00 P.M. EST

2. Click Apply, then Close.

3. Find the Sun.

4. Click on the General tab in the Object Information window and scroll down the information window.

5. At the bottom of the information window, you will find the seasonal information.

6. When does the summer solstice occur in 2010 C.E.? _____
(June 21st at 6:29 A.M. EST)

7. Close the file and do not save it, or continue with Exercise L.

Exercise L: Determining When the Summer Solstice Occurred in 52 C.E.

1. Run *TheSky* and set as follows:

   Date: March 1, 52
   Time: 8:00 A.M. EST

2. Find the Sun.

3. When did the summer solstice occur in 52 C.E.?
(June 24th at 2:43 A.M. EST)

4. Close the file and do not save it, or continue with Exercise M.

Exercise M: Determining When the Summer Solstice Occurred in 2735 B.C.E.

1. Run *TheSky* and set as follows:

   Date: July 4, −2734
   Time: 12:00 EST

2. Find the Sun.

3. When did the summer solstice occur in 2735 B.C.E.? _____

4. Close the file and do not save it, or continue with Exercise N.

Exercise N: Determining When the Summer Solstice Will Occur in 9999 C.E.

1. Run *TheSky* and set as follows:

   Date: September 15, 9999.
   Time: 15:00 EST.

2. Find the Sun.

3. When does the summer solstice occur in 9999 C.E.? _____

4. Close the file and do not save it.

Autumnal Equinox

Each day that follows the summer solstice, the Sun continues its eastward motion along the ecliptic. A few days after the summer solstice the Sun's apparent motion is both eastward and southward along the ecliptic. It becomes evident once again that the point where the ecliptic intersects the eastern horizon continues moving, but this time the motion is southward.

The Sun eventually crosses the celestial equator and moves from the Northern Hemisphere of the sky into the Southern Hemisphere. This occurs on or about September 21st each year and is known as the autumnal equinox. If the Sun happens to be located at the autumnal equinox, then this point in space becomes a point in time. That is, it is the first day of autumn in the Northern Hemisphere. The equatorial coordinates of the Sun on the celestial sphere at the time of the autumnal equinox are right ascension $12^H\ 00^M\ 00^S$ and declination $00°\ 00'\ 00''$. In other words, the Sun's location on the celestial sphere is precisely on the celestial equator once again.

Figure 8-16 displays the location of the autumnal equinox in *TheSky*. The view is from Ball State Observatory. The autumnal equinox is positioned on the local celestial meridian once again. The altitude of this point is about 50°, and its azimuth is 180°. At the bottom of the window, the right ascension coordinates are displayed.

Figure 8-16 Location of the autumnal equinox as shown in *TheSky*

If you toggle on the constellation boundaries, this point appears to be located near the western edge of the constellation of Virgo. In about 500 to 600 years from now, this point will move into the constellation of Leo and will take nearly 2000 years to move through it. This apparent motion is the result of Earth's precession. The sky window in Figure 8-16 may be viewed in the multimedia file named AutumnalEquinox.sky.

As Earth moves in its orbit from the summer solstice to the autumnal equinox, the days in the Northern Hemisphere become shorter and cooler and those in the Southern Hemisphere become longer and warmer. The apparent rising point of the Sun continues to change position along the eastern horizon. It appears to be moving southward now. To view the change of the Sun's position on the eastern horizon, complete Exercise O.

Exercise O: Observing Position of Sun's Rising Point

1. Run *TheSky*.

2. Open the multimedia file named Season2.sky.

3. Click on the ▶▶ button. The time step is 2 days.

4. After 6 seconds, click the 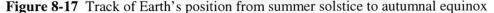 button and observe the position of the yellow dotted line on the horizon. Describe what happened. _____
_____

5. Close the file and do not save it.

The time sequence, in Exercise O, begins with the apparent sunrise on June 20, 2092, and continues through September 21, 2092, for an observer located at Ball State Observatory. Notice that the Sun immediately disappears below the eastern horizon. This is because the Sun appears to rise later each day from June until September. The ecliptic appears to move southward along the eastern horizon—watch the yellow dotted line again.

If Earth is viewed from space during this time interval, its axis of rotation moves from its alignment on the Sun to where it neither points toward or away from the Sun again. Earth has moved an additional 90° from the vernal equinox. The sunlight begins to shine less directly on the Northern Hemisphere and more so on the Southern Hemisphere.

The track of Earth's position during the time from the summer solstice to the autumnal equinox can be seen in Figure 8-17.

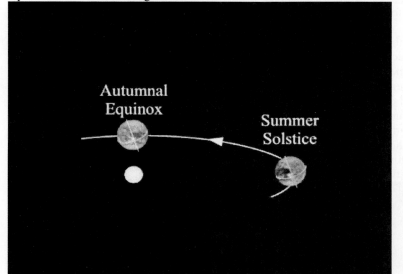

Figure 8-17 Track of Earth's position from summer solstice to autumnal equinox

As Earth continues its journey around the Sun, it does *not* lean toward or away from the Sun. Its orientation in space remains the same as before. The Sun shines directly onto Earth's equatorial region again and equally on both hemispheres, as shown in Figure 8-18.

If Earth is viewed from space at the time of the autumnal equinox, its axis of rotation once again does not point toward or away from the Sun. Earth now is exactly 180° in its orbit from the time of the vernal equinox. The lengths of days and nights are nearly equal at the time of the autumnal equinox. The autumnal equinox occurs at precisely 6:40 P.M. EST at Ball State Observatory on September 21, 2092. The Sun's equatorial coordinates on the celestial sphere at the time of the autumnal equinox are right ascension $12^H 00^M 00^S$ and declination $00° 00' 00''$. That is, the Sun is precisely on the celestial equator.

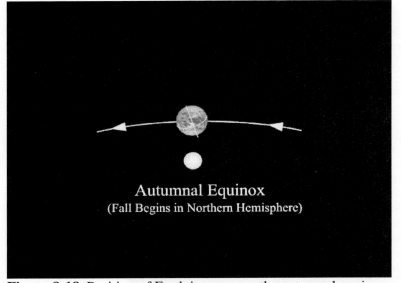

Figure 8-18 Position of Earth in space on the autumnal equinox

The position of the apparent rising point of the Sun once again changes throughout this time sequence. It reaches the East Point on the horizon on or about September 21st each year, and the Sun appears to rise due east as it did on the vernal equinox as was shown in Figure 8-5. At the beginning of this sequence, June 20th, the Sun appears to rise at 5:19 A.M. while at the end of the sequence, September 21st; it appears to rise at 6:28 A.M. EST.

The apparent sunrise for an observer located at Ball State Observatory on September 21, 2092, is illustrated in Figure 8-19. Notice that the equatorial grid is displayed once again. The Sun reaches the autumnal equinox point at precisely 6:23 P.M. EST on this date. The sky window again clearly shows the yellow dotted line, representing the ecliptic, intersecting the celestial equator near the East Point on the horizon.

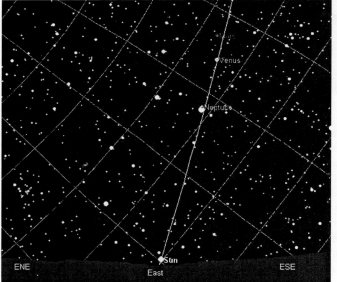

Figure 8-19 Apparent sunrise, September 21, 2092, from Ball State Observatory

If you travel to the geographic equator and observe the apparent sunrise there, the Sun once again appears to rise due east at that location. Figure 8-20 illustrates the apparent sunrise for an observer at the geographic equator. The equatorial grid is displayed in this sky window.

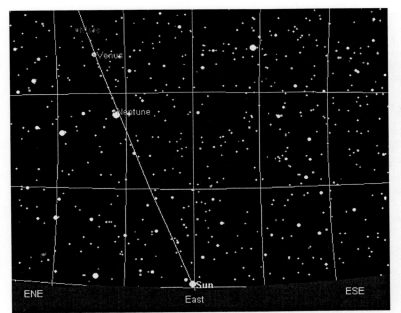

Figure 8-20 Apparent sunrise, September 21, 2092, at the geographic equator

At Ball State Observatory, the Sun's altitude constantly changes throughout the morning hours, and it eventually crosses the local celestial meridian. At midday, the Sun's altitude while crossing the local celestial meridian is the same as it was on the vernal equinox. The Sun's position in the sky looks exactly as it did in Figure 8-7.

Figure 8-21 shows the Sun's position in the sky at midday for an observer at Ball State Observatory at 12:34 P.M. EST. Notice that the horizon grid is displayed and that the Sun's altitude is actually 49° 53′ 56″ as found in the Object Information window for this date.

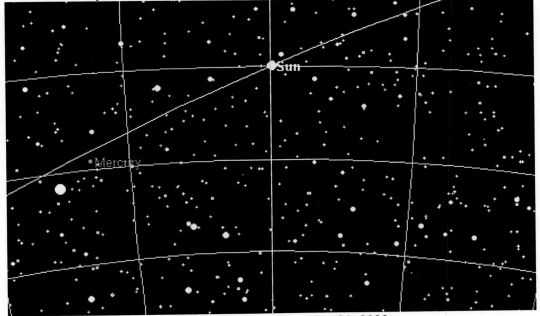

Figure 8-21 Altitude of the Sun at Midday, September 21, 2092, at Ball State Observatory

The position of the Sun for an observer at a different location on Earth on September 21, 2092, at midday is illustrated in Figure 8-22. Notice that the horizon grid is once again displayed. Where is the observer's location on Earth?

Figure 8-22 Altitude of the Sun at Midday, September 21, 2092, from an undisclosed location on Earth

The location of the observer in Figure 8-22 is, of course, the geographic equator.

The position of the Sun on the celestial sphere is directly over Earth's equator at the time of the autumnal equinox. The Sun's position in the sky on the autumnal equinox in September is shown in Figure 8-23. It is obvious that the Sun shines directly on both hemispheres in September, as it did in March.

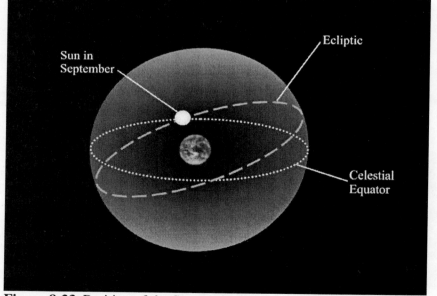

Figure 8-23 Position of the Sun on the celestial sphere on the autumnal equinox in September

Exercises P, Q, R, and S, which follow, are designed to help you determine when autumn occurs in the Northern Hemisphere for a particular year.

Exercise P: Determining When the Autumnal Equinox Will Occur in 2292 C.E.

1. Run *TheSky* and set as follows:

    Date: March 15, 2292
    Time: 10:00 A.M. EST

2. Click Apply, then Close.

3. Find the Sun and center it.

4. Click on the General tab in the Object Information window and Scroll down the information window.

5. When does the autumnal equinox occur in 2292 C.E.? _____
    (The answer is September 22nd at 3:55 A.M. EST)

6. Close the file and do not save it, or continue with Exercise Q.

Exercise Q: Determining When the Autumnal Equinox Occurred in 52 C.E.

1. Run *TheSky* and set as follows:

    Date: March 1, 52 C.E.
    Time: 8:00 A.M. EST

2. Find the Sun and center it.

3. When did the autumnal equinox occur in 52 C.E.? _____
    (The answer is September 24th at 2:31 P.M. EST)

4. Close the file and do not save it, or continue with Exercise R.

Exercise R: Determining When the Autumnal Equinox Occurred in 2735 B.C.E.

1. Run *TheSky* and set as follows:

    Date: July 4, −2734
    Time: 12:00 EST

2. Find the Sun and center it.

3. When did the autumnal equinox occur in 2735 B.C.E.? _____

4. Close the file and do not save it, or continue with Exercise S.

Exercise S: Determining When the Autumnal Equinox Will Occur in 9999 C.E.

1. Run *TheSky* and set as follows:

   Date: September 15, 9999
   Time: 3:00 P.M. EST

2. Find the Sun and center it.

3. When does the autumnal equinox occur in 9999 C.E.? _____

4. Close the file and do not save it.

Winter Solstice

The Sun crosses the celestial equator on the autumnal equinox and begins to move into the Southern Hemisphere of the sky. Each day that follows the autumnal equinox, the Sun's motion carries it farther and farther into the Southern Hemisphere of the sky. Several days after the autumnal equinox the Sun's motion continues to be eastward and southward.

If Earth is viewed from space during this time, the southern half of its axis of rotation begins to align itself on the Sun. Earth has now moved 90° in its orbit from the autumnal equinox and 270° from the vernal equinox. Sunlight begins to shine more directly on the Southern Hemisphere and less so on the Northern Hemisphere. The days in the Northern Hemisphere become shorter and cooler, and those in the Southern Hemisphere become longer and warmer.

The track of Earth's position during the time from the autumnal equinox to the winter solstice can be seen in Figure 8-24. While Earth moves around the Sun during this time period, it does *not* lean away from the Sun.

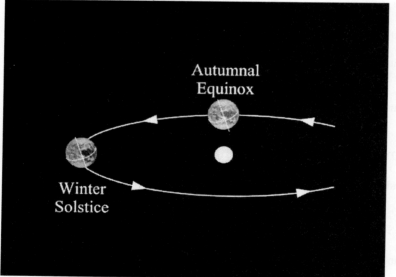

Figure 8-24 Track of Earth's position from autumnal equinox to winter solstice

Its orientation in space has remained the same way as before. The position of Earth in space at the time of the winter solstice is shown in Figure 8-25.

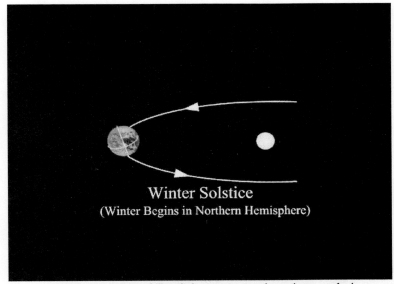

Figure 8-25 Position of Earth in space on the winter solstice

As before, it becomes apparent once again that the rising point of the Sun begins moving farther South on the eastern horizon. To view this change of position in the Sun's rising point on the eastern horizon, complete Exercise T.

Exercise T: Observing Position of Sun's Rising Point

1. Run *TheSky*.

2. Open the multimedia file named Season3.sky.

3. Click on the [▶▶] button. The time step is 2 days.

4. After 6 seconds, click the [■] button and observe the position of the yellow dotted line on the horizon. Describe what happened. _____

   _____

5. Close the file and do not save it.

The time sequence, in Exercise T, begins at the apparent sunrise on September 21, 2092, and continues to December 20, 2092, for an observer located at Ball State Observatory. Almost immediately the Sun disappears below the eastern horizon. This is because the Sun appears to rise later each day from September until December. The ecliptic appears to continue moving southward along the eastern horizon—watch the yellow dotted line along the eastern horizon.

As Earth moves in its orbit from the autumnal equinox throughout this time interval, the intersection of the ecliptic and the horizon continues to change its position along the eastern horizon. The apparent rising point of the Sun changes its position on the eastern horizon throughout this period of time. It is now moving southward along the horizon. It reaches its most southerly position on the horizon on or about December 20th each year.

At the most southerly point on the eastern horizon, the Sun appears to stop its southward motion. This is the other solstice point. That is, the Sun "stands still" on the horizon once again. The Sun has reached the winter solstice point. At the beginning of the sequence in Season3.sky, September 20th, the Sun appears to rise at 6:29 A.M. EST, and at the end of the sequence, December 20th, the Sun appears to rise at 8:00 A.M. EST. During this time interval, Earth does *not* lean away from the Sun. Its orientation in space has always been this way as before!

The apparent sunrise is shown in Figure 8-26 for an observer located at Ball State Observatory on December 20, 2092, at 8:00 A.M. EST. The yellow dotted line, representing the Sun's apparent path, is located approximately 31° south of due east at this location. The winter solstice occurs at precisely at 4:34 P.M. EST on this date. Notice that the equatorial grid is displayed. The Sun's equatorial coordinates on the celestial sphere at the time of the winter solstice are right ascension $18^H 00^M 00^S$ and declination $-23° 26'$. This means, of course, that the Sun is an additional 90° from the autumnal equinox and 270° from the vernal equinox.

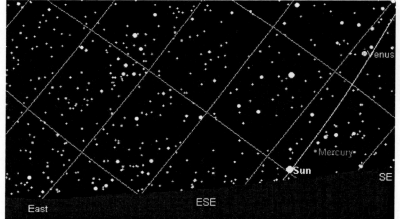

Figure 8-26 Apparent sunrise on December 20, 2092, at Ball State Observatory

If an observer is located at the geographic equator, the apparent rising point of the Sun is exactly 23° 26' south of the East Point. The apparent sunrise for an observer located at the geographic equator is displayed in Figure 8-27. Once again the equatorial grid is displayed.

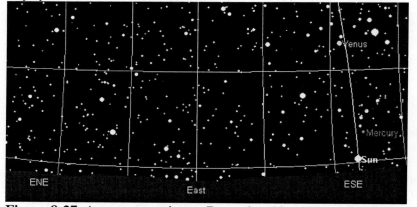

Figure 8-27 Apparent sunrise on December 20, 2092, at the geographic equator

On the winter solstice, the Sun has the lowest altitude at midday for any observer located in the Northern Hemisphere. Figure 8-28 shows the position of the Sun at midday for an observer located at Ball State Observatory on December 20, 2092. Notice that the horizon grid is displayed. The Sun has an altitude of about 26° on December 20th at Ball State Observatory. You can look in the Object Information window to find the altitude of the Sun at midday for this date. Is it about 26° for this date at this location? The location of the winter solstice may be viewed in the multimedia file named WinterSolstice.sky.

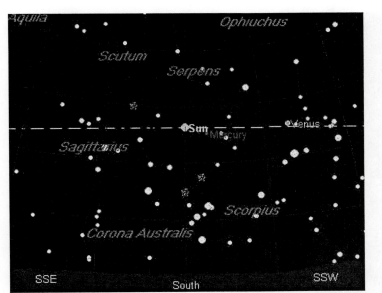

Figure 8-28 Altitude of the Sun, at midday, December 20, 2092, at Ball State Observatory

For any observer located on the geographic equator, the Sun's position at midday is exactly 23° 26′ south of their zenith. In fact, the Sun is located at the zenith of observers who live on the 23° 26′ south latitude circle (Tropic of Capricorn) at midday. To view the Sun from the Tropic of Capricorn on December 20, 2092, complete Exercise U.

Exercise U: Observing the Sun at Midday on the Tropic of Capricorn

1. Run *TheSky*.

2. Open the multimedia file named Season4.sky.

3. Set time: 12:39 P.M. EST.

4. Set Latitude: 23° 26′ S.

5. Turn on the horizon grid.

6. Find the Sun and center it.

7. Your computer screen should look like the one displayed in Figure 8-29.

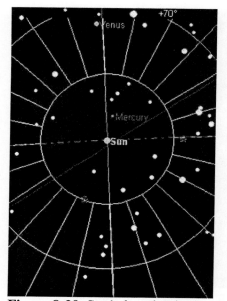

Figure 8-29 Sun's location in the sky at midday on the Tropic of Capricorn

8. What is the altitude of the Sun? _____ °

9. Close the file and do not save it.

Figure 8-30 illustrates the position of the Sun on the celestial sphere on the winter solstice. It is obvious that sunlight shines most directly on the Southern Hemisphere in December.

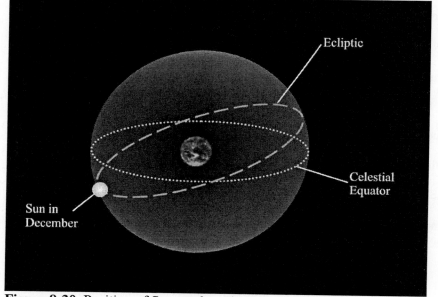

Figure 8-30 Position of Sun on the celestial sphere on the winter solstice in December

Exercises V, W, X, and Y are designed to help you to determine when winter occurs for a particular year in the Northern Hemisphere.

Exercise V: Determining When the Winter Solstice Will Occur in 2292 C.E.

1. Run *TheSky* and set as follows:

   Date: November 15, 2292
   Time: 10:00 A.M. EST

2. Click Apply, then Close.

3. Find the Sun and center it.

4. Click on the General tab in the Object Information window and Scroll down the information window.

5. When does the winter solstice occur in 2292 C.E.? _____
   (The answer is December 21st at 5:21 A.M. EST)

6. Close the file and do not save it, or continue with Exercise W.

Exercise W: Determining When the Winter Solstice Occurred in 52 C.E.

1. Run *TheSky* and set as follows:

   Date: October 12, 52 C.E.
   Time: 8:00 A.M. EST

2. Find the Sun and center it.

3. When did the winter solstice occur in 52 C.E.? _____
   (The answer is December 22nd at 7:42 A.M. EST)

4. Close the file and do not save it, or continue with Exercise X.

Exercise X: Determining When the Winter Solstice Occurred in 2735 B.C.E.

1. Run *TheSky* and set as follows:

   Date: April 1, −2734
   Time: 12:00 EST

2. Find the Sun and center it.

3. When does the winter solstice occur in 2735 B.C.E.? _____

4. Close the file and do not save it, or continue with Exercise Y.

Exercise Y: Determining When the Winter Solstice Will Occur in 9999 C.E.

1. Run *TheSky* and set as follows:

   Date: September 29, 9999
   Time: 3:00 P.M. EST

2. Find the Sun and center it.

3. When does the winter solstice occur in 9999 C.E.? _____

4. Close the file and do not save it, or continue with Exercise Z.

Each day that follows the winter solstice, the Sun continues to move eastward in the sky. A few days after the winter solstice the Sun's motion is both eastward and northward along the ecliptic. The days become longer and warmer in the Northern Hemisphere and those in the Southern Hemisphere become shorter and cooler.

It becomes apparent that the rising point of the Sun is moving farther north on the eastern horizon. To view the rising point of the Sun change its position on the eastern horizon during this time interval, complete Exercise Z.

Exercise Z: Observing Position of Sun's Rising Point

1. Run *TheSky*.

2. Open the multimedia file named Season4.sky.

3. Click on the ▶▶ button. The time step is 2 days.

4. After 5 seconds, click the ■ button and observe the position of the yellow dotted line on the horizon. Describe what happened. _____
   _____

5. Close the file and do not save it.

The time sequence in Exercise Z begins at the apparent sunrise on December 21, 2092, and continues through March 19, 2093, for an observer at Ball State Observatory. Right away the observer notices that the Sun's altitude once again gets higher in the sky. This is because the Sun appears to rise earlier each day from December until March. Another thing to notice is that the ecliptic appears to move northward along the eastern horizon. That is, the apparent rising point of the Sun continues to change throughout this time sequence also.

The Sun appears to be heading back toward the celestial equator. It will take three months for it to reach the celestial equator and the vernal equinox once again. The Sun finally reaches the East Point on the horizon on March 19, 2093. At the beginning of the sequence, on December 20, 2092, the Sun appears to rise at 8:00 A.M. EST whereas at the end of the sequence, on March 19, 2093, the Sun appears to rise at 6:45 A.M. EST for an observer at Ball State Observatory. The Sun reaches the vernal equinox point at precisely 3:37 P.M. EST on this date. The exact location of the Sun on the celestial sphere is once again at right ascension $00^H\ 00^M\ 00^S$ and declination $00°\ 00'\ 00''$. The apparent rising point of the Sun is shown in Figure 8-31 for an observer located at Ball State Observatory on March 19, 2093.

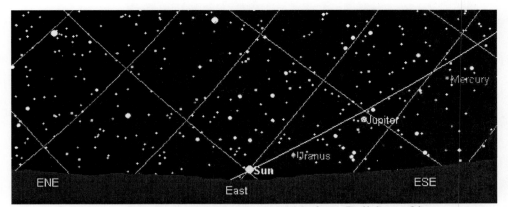

Figure 8-31 Apparent sunrise on March 19, 2093, from Ball State Observatory

If the Earth is viewed from space during this interval of time, the southern half of the its axis of rotation begins to move away from its alignment on the Sun. The Earth has moved 90° in its orbit from the winter solstice and 360° from the vernal equinox point where it started in 2092. The Sun has finally come full circle to where we began our journey and our discussion of the seasons.

The track of Earth's position during the time from the winter solstice to the vernal equinox can be seen in Figure 8-32. As Earth continues its motion around the Sun, once again it does *not* lean away from the Sun; its orientation in space has remained the same as it has throughout the year.

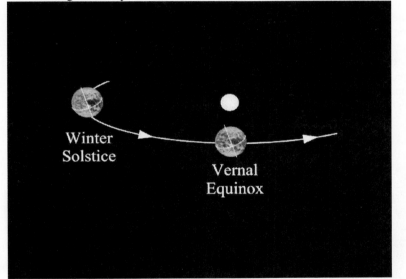

Figure 8-32 Track of Earth's position from winter solstice to vernal equinox

The Sun once again shines directly onto the Earth's equatorial region and equally on both hemispheres, as shown in Figure 8-33. The lengths of days and nights are nearly equal, as they were when we started a year ago. It is springtime once more in the Northern Hemisphere! And so it goes, year after year, after year.

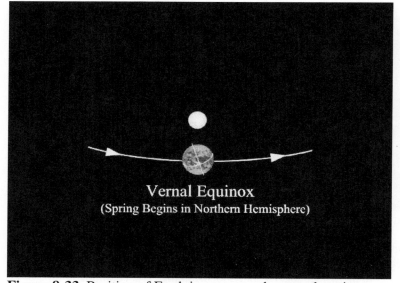

Figure 8-33 Position of Earth in space on the vernal equinox

Observers who will watch the sky over the decades between 2001 and 2092 will notice the same constellations appear, disappear, and reappear at precisely the same time of year, each year. Even though the dates (days) of the seasons change throughout this period of time, it soon becomes apparent that the seasons occur when the Sun unequivocally reaches the same point on the celestial sphere each year. The appearance of these seasonal constellations was the basis for ancient time systems and calendars.

One such ancient calendar is Stonehenge, perhaps the most famous of the ancient calendars (see Figure 8-34). If you stand in the center of these stones, you can accurately ascertain the date by observing the position of the Sun's apparent rising point on the eastern horizon. Stonehenge, like many other assemblies of megaliths in England, was built by prehistoric Celts thousands of years ago. Most such calendars are as accurate today as they were in ancient times.

Figure 8-34 Stonehenge near Salisbury, England
(Photo Courtesy of Elle Smith)

The North American continent is no stranger to ancient calendars. Native Americans apparently built calendars similar to Stonehenge on this continent too. Their scale pales in comparison to the megalith calendars in Europe, but they were quite accurate nonetheless.

Figure 8-35 shows the Bighorn Medicine Wheel on Medicine Mountain, located west of Sheridan, Wyoming, built about 7500 years ago.

Figure 8-35 Bighorn Medicine Wheel near Sheridan, Wyoming, (Photo Courtesy of U.S. Forest Service).

Modern calendars have incorporated the motion of precession. The seasons we recognize today *will* begin on about the same dates in the future as they do in this epoch. And the holidays we celebrate today will probably be celebrated on the same dates in the future. That is, most people will celebrate Christmas on December 25th and *Easter* will be (as it is today) the *first* Sunday *after* the *first* full Moon *after* the *vernal equinox* each year.

The nighttime sky, however, will tell a different story because of Earth's precession. The constellations you see today in your summer sky will be some of the ones people will see in the winter sky thousands of years from now, while those in your winter sky today will be some of the summer constellations seen in 13,000 years. Halfway through the precessional cycle of Earth, the bright star Vega will become our North Star, as it was in 11,000 B.C.E. *TheSky* lets you view the projected results of Earth's precession.

As you become more and more proficient using *TheSky*, it is possible to take trips into the future or back into the ancient past. It is left as an exercise for the reader to make these trips and to explore the capabilities of *TheSky*. Who knows what ancient mysteries *you* might discover or uncover in pursuing time travel explorations using *TheSky*? *TheSky* lets you set the date of the sky window from 4713 B.C.E. to 10,000 C.E. This involves a lot of events in human history.

Chapter 8

TheSky Review Exercises

Answer these questions using *TheSky* software

TheSky Review Exercise 1: Determining the Date of the Vernal Equinox

1. On what date does the vernal equinox occur in the year 2025 C.E.? _____

2. What time does the event occur in Question 1 in Washington, D.C.? _____:_____

3. What time does the event occur in Question 1 in Waterloo, IA? _____: _____

4. What time does the event occur in Question 1 in Seattle, Washington? _____:_____

5. What time does the event occur in Question 1 in Cleveland, Ohio? _____:_____

6. In what constellation does the Sun appear to be located in Question 1?
    _____

7. In what constellation does the Sun appear to be located in the year 2825 C.E.?
    _____

8. Does the Sun appear to be located in the same constellation in Questions 6 and 7? _____
    If not, why not? _____

TheSky Review Exercise 2: Determining the Date of the Summer Solstice

1. On what date does the summer solstice occur in the year 6500 C.E.? _____

2. What time does the event in Question 1 occur in Greenwich, England? _____:_____

3. What time does the event in Question 1 occur in New York City, NY? _____:_____

4. What time does the event in Question 1 occur at Ball State Observatory? ____:_____

5. What time does the event in Question 1 occur in Los Angeles, CA? _____:_____

6. What time does the event in Question 1 occur in Sydney, Australia? _____:_____

7. In what constellation does the Sun appear to be located on the date in Question 1?
    _____

8. In what constellation doe the Sun appear to be located in 4500 C.E.?
    _____

9. Does the Sun appear to be located in the same constellation in Questions 7 and 8? _____
    If not, why not? _____

TheSky Review Exercise 3: Determining the Date of the Autumnal Equinox

1. On what date does the autumnal equinox occur in the year 2215 C.E.? _____

2. What time does the event in Question 1 occur in Greenwich, England? _____:_____

3. What time does the event in Question 1 occur in Columbus, GA? _____:_____

4. What time does the event in Question 1 occur in Miami, FL? _____:_____

5. What time does the event in Question 1 occur in Cleveland, OH? _____:_____

6. In what constellation does the Sun appear to be located on the date in Question 1?
   _____

7. In what constellation does the Sun appear to be located in the year 5515 C.E.?
   _____

8. Does the Sun appear to be located in the same constellation in Questions 7 and 8?
   _____ If not, why not? _____

TheSky Review Exercise 4: Determining the Date of the Winter Solstice

1. On what date does the winter solstice occur in the year 8592 C.E.? _____

2. At what time does the event in Question 1 occur in Atlanta, GA? _____:_____

3. At what time does the event in Question 1 occur in Athens, Greece? _____:_____

4. At what time does the event in Question 1 occur in Sydney, Australia? _____:_____

5. In what constellation does the Sun appear to be located in Question 1?

6. In what constellation does the Sun appear to be located in the year 1492 C.E.?

7. Does the Sun appear to be located in the same constellations in Questions 5 and 6?
   ____ If not, why not? _____

Chapter 8

TheSky Review Questions

Answer these following review questions concerning the material in the chapter.

1. What is the name of the point in space where the Sun appears to cross the celestial equator from South to North? _____ _____

2. On what date does the Sun appear to cross the point on the celestial equator in Question 1? _____

3. What are the Sun's right ascension and declination at the vernal equinox?

    R.A. = _____^{H}_____^{M} Declination = __ _____$^\circ$_____$'$

4. On what date does the Sun appear to be 23½° North of the celestial equator?

    _____

5. What is the name of the point on the celestial sphere in Question 4?

    _____ _____

6. What are the right ascension and declination of the Sun on approximately June 21st?

    R.A. = ____^{H}____^{M} Declination = __ _____$^\circ$_____$'$

7. On what date does the Sun appear to cross the celestial equator from North to South?

    _____

8. What is the name of the point on the celestial sphere in Question 7?

    _____ _____

9. What are the right ascension and declination of the Sun at the time of the autumnal equinox?

    R.A. = ____^{H}____^{M} Declination = __ _____$^\circ$_____$'$

10. On what date does the Sun appear to be 23½° South of the celestial equator?

    _____

11. What is the name of the point on the celestial sphere in Question 10?

    _____ _____

12. What are the right ascension and declination of the Sun on approximately December 21st?

    R.A. = _____^{H} _____^{M} Declination = __ _____° _____'

Chapter 9

Phases and Eclipses in *TheSky*

Motion of the Moon

One of the most obvious objects that people notice in the night sky is the Moon. For millennia humans have observed it and constructed calendars by it. It really is something that is straightforward to observe. It is an easy task to observe the Moon—only time must be taken to do it.

The ancient Greeks considered the Moon a planet because, it moved with respect to the constellations. However, today we realize that it is not a planet but merely a satellite of Earth. Like, the Sun, it moves on a well-defined path in the sky near the ecliptic. Its path, however, is tilted 5° 8′ 43″ with respect to the Sun's apparent path. As the Moon revolves around Earth, its declination varies from the declination of the Sun in the sky during the course of a month. Sometimes the Moon is above the ecliptic, and sometimes it is below.

The Moon's motion among the stars is easy to observe, because it moves its own diameter every hour. Its eastward motion amounts to approximately ½° per hour, which is approximately 13° each day relative to the stars. In fact, its angular motion is by far greater than any other astronomical object observed in the sky. In Exercise A, the path and motion of the Moon is illustrated with respect to the stars and the ecliptic.

Exercise A: Motion of the Moon

1. Run *TheSky*.

2. Open the multimedia file named MoonMotion.sky.

3. Click the ⏭ button.

4. Observe the Moon's motion relative to the background stars. The motion is in 1-hour time intervals.

5. Click the Display Right button (➡) on the Orientation toolbar to follow the Moon through a complete cycle in the sky.

6. Notice that the Moon not only changes its position in the sky but also goes through a phasing process.

7. Change the time increment from 1 hour to 1 sidereal day.

8. Observe how much the Moon moves each day.

9. Close the file and do not save it.

The Moon takes about a month to complete one circuit of the sky. This is the Moon's sidereal period. The actual time is 27.322 days. During this time, as the Moon moves counterclockwise around Earth its position with respect to the Sun constantly changes. As a result of this motion, different amounts of the lunar surface are illuminated and become visible from Earth's surface. In other words, as observed from Earth, the Moon appears to go through a series of phases.

Phases of the Moon

The Moon's phasing period has a time interval that is longer than its sidereal period. The phase period is a time interval *with respect to* the *Sun*, not to the stars. This period of time is also about a month. Actually, it takes the Moon 29.531 days to go through a complete phasing cycle. This period of time is known as the Moon's synodic period. Timekeepers have been aware of this cycle for hundreds or perhaps thousands of years.

Figure 9-1 is a schematic diagram of the Moon's orbit around Earth as viewed from some distance away in space. The motion of the Moon about Earth in this figure is counterclockwise as viewed from above the Earth-Moon System.

The arrows at the right-hand side in Figure 9-1 represent sunlight passing through the Earth-Moon System. The Sun's distance is so great in this figure that its rays pass through the system essentially parallel to one another. Earth and Moon are equally illuminated on the half toward the Sun. The lighter areas on the right-hand side of Earth and Moon represent the day side while the dark shaded areas opposite the Sun's rays represent the night side on Earth and Moon.

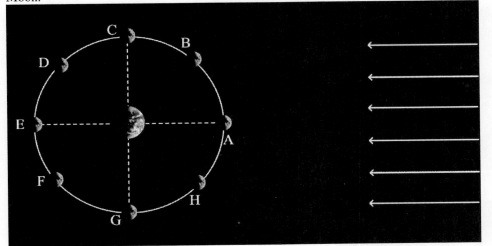

Figure 9-1 Earth-Moon System

New Moon

As viewed from Earth, the Moon is aligned with the Sun at point A. If the Moon's angular separation (elongation) is measured from the Sun at this point, it is 0°. In other words, the right ascensions of the Sun and Moon are exactly the same when the Moon is located at point A. On carefully inspecting Figure 9-1, you will notice that at point A the Moon's night side faces Earth. This means that *none* of the Moon's illuminated surface can be seen from Earth's surface. The phase of the Moon at this point is called a new Moon. The Moon appears to rise and set with the Sun and is not visible to us on Earth, because it is above the horizon with the Sun in the day sky.

Figure 9-2 displays *TheSky's* view of a new Moon as viewed by an observer from Ball State Observatory on May 27, 2006, at 01:33 A.M. EST. In actuality, you could not see the Moon when its phase is a new Moon. However, Figure 9-3 illustrates how the Moon should look when viewed through a telescope on this date. Because the Moon is between Earth and the Sun its night side is facing Earth and is not visible.

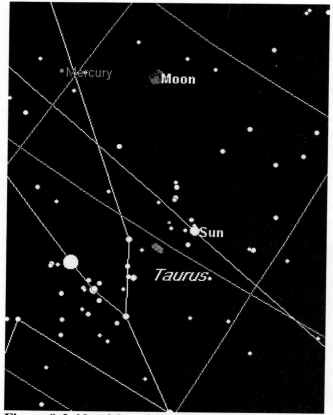

Figure 9-2 New Moon in Taurus on May 27, 2006

Figure 9-3 Appearance of a new Moon phase in Earth's sky (not visible)

Figure 9-4 shows the Object Information window for the Moon on May 27th. *TheSky*, rather than naming lunar phases, indicates a percentage of the surface of the Moon that is visible from Earth. It also indicates the angular separation between the Sun and Moon. This angle indicates how far apart the Sun and Moon appear to be from each other in the sky. On this date the Moon is approximately 3° north of the Sun. However, when you use the Find command to locate the Sun on May 27th at 01:37, the right ascensions of the Sun and Moon are nearly the same.

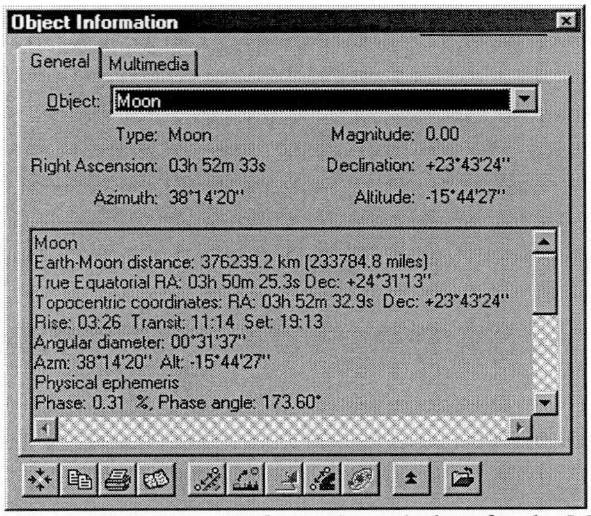

```
Object Information                                    [x]
 General  Multimedia
  Object: Moon                                         [v]
              Type: Moon              Magnitude: 0.00
  Right Ascension: 03h 52m 33s    Declination: +23°43'24"
           Azimuth: 38°14'20"         Altitude: -15°44'27"
  ┌─────────────────────────────────────────────────┐
  │ Moon                                          [▲] │
  │ Earth-Moon distance: 376239.2 km (233784.8 miles) │
  │ True Equatorial RA: 03h 50m 25.3s Dec: +24°31'13" │
  │ Topocentric coordinates: RA: 03h 52m 32.9s Dec: +23°43'24" │
  │ Rise: 03:26  Transit: 11:14  Set: 19:13           │
  │ Angular diameter: 00°31'37"                       │
  │ Azm: 38°14'20"  Alt: -15°44'27"                   │
  │ Physical ephemeris                                │
  │ Phase: 0.31 %, Phase angle: 173.60°          [▼] │
  │ [◄]                                          [►] │
  └─────────────────────────────────────────────────┘
```

Figure 9-4 Object Information window for the Moon on May 27th

If you would like to view the sky window displayed in Figure 9-2, then follow the instructions provided in Exercise B.

Exercise B: Observing the New Phase of the Moon on May 27, 2006

1. Run *TheSky*.

2. Open the multimedia file named NewMoon.sky.

3. Click on the Sun and record its coordinates here.

   R.A. = _____ ^{H} _____ ^{M} _____ ^{S} Declination = __ _____ $^\circ$ _____ $'$ _____ $''$

4. Find the Moon to display the Object Information window.

5. Center ([⊹]) the Moon in the sky window.

6. Find the Moon's coordinates and record them here.

   R.A. = _____ ^{H} _____ ^{M} _____ ^{S} Declination = __ _____ $^\circ$ _____ $'$ _____ $''$

7. Scroll down the tab on the right side of the window to find the Moon's phase or the amount of its surface that is illuminated.

   Is the phase = 0.0%? _____. What is the phase? _____%

8. Compare the Moon's right ascension to that of the Sun. Are they about the same? _____. In other words, is the difference in right ascension about 0^H? _____.

9. Close the file and do not save it.

   The Sun and Moon are both seen in this window, even though the Moon is actually not visible from Earth. The phase that is given is 0.16%. This means that because the Moon is 3° above the Sun, this much of the Moon might be seen from Earth. However, because the Moon is in the day sky on this date it is not visible at all from Earth's surface. It is in the direction of the Sun and has its dark side toward Earth. The Sun is too bright to be able to see the Moon. Do not try looking for the Moon when it is so close to the Sun!

Figure 9-5 portrays the lunar phases as they appear from Earth's surface. In this figure, "phase A," represents the new Moon phase and is displayed as a totally black circle at the left side of the diagram. This phase corresponds to the Moon's position at point A in Figure 9-1. The lunar cycle begins with the Moon aligned on the Sun at point A, in Figure 9-1, or when its phase is a new Moon. It will take one synodic month ($29\frac{1}{2}^D$) for the Moon to move from point A in its orbit to this point again or from one new Moon to the next new Moon. These lunar images were rendered using *TheSky*.

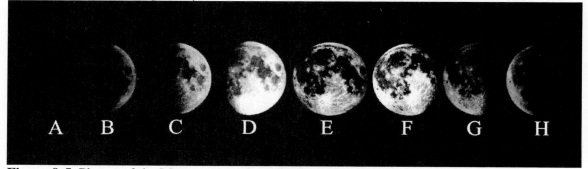

Figure 9-5 Phases of the Moon as seen from Earth's surface

First Quarter Moon

As the Moon moves counterclockwise around Earth in Figure 9-1, eastward in the sky as viewed from Earth, more and more of its illuminated surface becomes visible. The Moon eventually reaches point C in its orbit. It is now 90° east of the Sun. The Moon has moved one-quarter of the distance around its orbit at this point and is called a *first quarter phase*. The right ascension of the Moon is nearly 6^H greater (east) than the Sun's right ascension when the Moon is located at this point.

The time that it has taken for the Moon to move this far in its orbit is a little over a week (7.38 days). When the phase of the Moon is a first quarter, it appears to rise and set about 6 hours after the Sun. It is visible in the sky from about midday until just about midnight. From Earth, the Moon appears as a half circle or half-moon in the sky. It is the right half of the Moon that is seen. Figure 9-5, "phase C" represents the first quarter phase, and as you can see it appears as a half circle with the right half illuminated.

Figure 9-6 represents *TheSky's* rendition of a first quarter Moon seen in Leo on June 3, 2006, at 17:47 as viewed from Ball State Observatory.

Figure 9-6 First quarter phase of the Moon in Leo on June 3, 2006

Figure 9-7 is an image rendered in *TheSky* to show the appearance of a first quarter phase of the Moon as it looks when viewed through a telescope on this date. It is a zoomed-in view of the Moon in Figure 9-6. Notice that it resembles "phase C" in Figure 9-5.

Figure 9-7 Appearance of a first quarter phase of the Moon as seen in *TheSky*

If you like to view the sky window displayed in Figure 9-6, then follow the instructions provided in Exercise C.

Exercise C: Observing the First Quarter Phase of the Moon on June 3, 2006

1. Run *TheSky*.

2. Open the multimedia file named FirstQuarter.sky.

3. Click on the Sun and record its coordinates here.

   R.A. = _____ᴴ _____ᴹ _____ˢ   Declination = __ _____° _____′ _____″

4. Find the Moon to display the Object Information window.

5. Center () the Moon in the sky window.

6. Is the Phase about 50%? _____

7. What is the Moon and Sun's angular separation? _____° Is it about 90°? _____

8. Find the Moon's coordinates and record them here.

   R.A. = _____ᴴ _____ᴹ _____ˢ   Declination = __ _____° _____′ _____″

9. Compare the Moon's right ascension to that of the Sun's. Sun _____ᴴ Moon _____ᴴ
   Is the difference about 6ᴴ? _____

10. Use the Zoom button to observe the Moon's phase up close as it would appear if in a telescope.

11. Close the file and do not save it.

Full Moon

As the Moon continues its counterclockwise journey around Earth, it travels from point C to point E. By the time the Moon reaches point E, it has traveled halfway (two-quarters) around its orbit. At point E, the Moon's position in its orbit is exactly 180° from the Sun; that is, its position in the sky is directly opposite the Sun. The right ascension of the Moon at this point is exactly 12^H greater (east) than the Sun's right ascension. The time it has taken the Moon to reach this point in its orbit is a little over two weeks (14.77 days).

From Earth, the Moon now appears as a full circle. Whenever the Moon is located at this point in its orbit, its phase is called a *full phase*. In Figure 9-5, "phase E" represents the full phase and is displayed as a fully illuminated Moon. When the Moon is a full Moon it appears to rise when the Sun appears to set and appears to set when the Sun appears to rise. It is visible in Earth's sky throughout the entire night.

Figure 9-8 displays the full phase of the Moon in Ophiuchus on June 11, 2006 at 13:34 as viewed by an observer from Ball State Observatory.

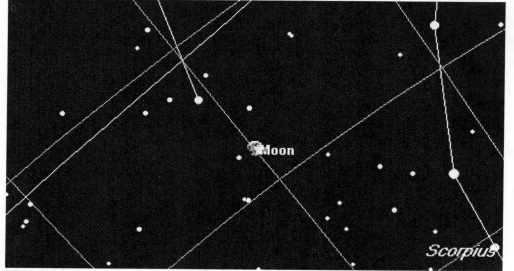

Figure 9-8 Full Moon phase as seen in Ophiuchus on June 11, 2006

Figure 9-9 displays *TheSky's* rendition of what a full Moon phase looks like when viewed with a telescope on this date. It is a zoomed-in view of the Moon in Figure 9-8. If you look at Figure 9-5, you will notice that it does resemble "phase E."

Figure 9-9 Appearance of a full phase of the Moon as seen in *TheSky*

If you would like to view the sky window displayed in Figure 9-8, then follow the instructions provided in Exercise D.

Exercise D: Observing the Full Phase of the Moon on June 11, 2006

1. Run *TheSky*.

2. Open the multimedia file named FullMoon.sky.

3. Click on the Sun and record its equatorial coordinates here.

    R.A. = _____ H _____ M _____ S   Declination = __ _____ ° _____ ' _____ "

4. Find the Moon to display the Object Information window.

5. Center (⊡) the Moon in the sky window.

6. Is the Phase about 100%? _____

7. What is the Moon and Sun's angular separation? _____°. Is it about 180°? _____

8. Find the Moon's equatorial coordinates and record them here.

    R.A. = _____ H _____ M _____ S   Declination = __ _____ ° _____ ' _____ "

9. Compare the Moon's right ascension to that of the Sun. Sun _____ H Moon _____ H
   Is the difference about 12^H? _____

10. Use the Zoom button to observe the Moon up close as in a telescope.

11. Close the file and do not save it.

Last Quarter Moon

The Moon continues to move in its orbit from point E to point G in Figure 9-1. When it reaches point G, the Moon is once again located 90° from the Sun. It is, however, now 90° west of the Sun or 270° from the point where the cycle began (new Moon). The Moon has moved three-quarters of the distance around its orbit to reach this point. The Moon's phase at point G is again called a quarter phase.

Its right ascension is now about 18^H greater (east) than the Sun's right ascension at this point. Or, it is 6^H less than the Sun's right ascension because it is west of the Sun. The time it has taken the Moon to move this far in its orbit is a little over three weeks (22.15 days). The Moon's phase at point G was once called the 3rd quarter phase but is now known as a *last quarter phase* of the Moon. When the Moon is a last quarter Moon, it appears to rise and set approximately 6 hours before the Sun (remember it is west). It is visible in Earth's sky from about midnight until approximately midday.

From Earth, the Moon once again appears as a half circle or a half-moon in the sky. But it is the *left half* of the Moon that is seen not the *right half* like the first quarter phase. The reason for this is that the Moon is now located west of the Sun. In Figure 9-5, "phase G" represents the last quarter phase in this figure and is displayed as a half circle with the left half of the Moon illuminated. Figure 9-10 displays the last quarter phase of the Moon in Pisces on June 18, 2006, at 09:25 for an observer located at Ball State Observatory.

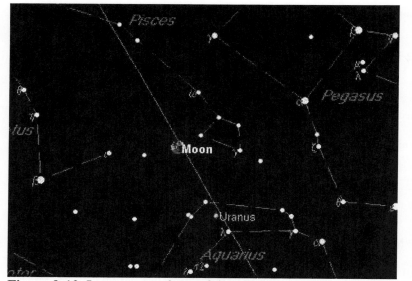

Figure 9-10 Last quarter phase of the Moon in Pisces on June 18, 2006

Figure 9-11 is *TheSky's* rendition of how a last quarter phase of the Moon would look if viewed through a telescope on this date. It is a zoomed in view of the Moon in Figure 9-10. Notice that it looks like "phase G" in Figure 9-5.

Figure 9-11 Appearance of a last quarter phase of the Moon as seen in *TheSky*

If you would like to view the sky window displayed in Figure 9-10, simply follow the instructions provided in Exercise E.

Exercise E: Observing the Last Quarter Phase of the Moon on June 18, 2006

1. Run *TheSky*.

2. Open the multimedia file named LastQuarter.sky.

3. Click on the Sun and record its equatorial coordinates here.

    R.A. = _____ ^H _____ ^M _____ ^S   Declination = __ _____ ° _____ ' _____ "

4. Find the Moon to display the Object Information window.

5. Center ( ) the Moon in the sky window.

6. Is the Phase about 50%? _____

7. What is the Moon and Sun's angular separation? _____° Is it about 90°? _____

8. Find the Moon's equatorial coordinates and record them here.

    R.A. = _____^{H} _____^{M} _____^{S} Declination = __ _____° _____′ _____″

9. Compare the Moon's right ascension to that of the Sun. Sun _____^{H} Moon _____^{H}
   Is the difference about 18^H? _____

10. Use the Zoom button to observe the Moon up close as in a telescope.

11. Close the file and do not save it.

   The Moon finally moves from point G to point A in Figure 9-1, thus completing one lunar cycle about Earth. The total time it has taken the Moon to complete this cycle is 29.531 days with respect to the Sun. This is the Moon's *synodic* or *phase period*! At this point the cycle starts over again.

Other Phases of the Moon

The phases discussed so far are those which have well-defined measured positions in the Moon's orbit. The angles of these phases in the Moon's orbit with respect to Earth and the Sun are precisely known. That is, the angles, or one might say elongations are 0°, 90°, 180°, and 270° (new, first quarter, full, last quarter). However, during the time that it has taken the Moon to complete one phase cycle around Earth, it has also exhibited a number of other phases in Earth's sky.

   For example, as the Moon moves from point A to point C the sunlit portion of the Moon's surface gradually becomes more visible to observers on Earth. The Moon appears to go through a series of crescent phases in Earth's sky, each larger than the previous night, during its phase cycle. The illuminated areas are visible on the right edge of the Moon as viewed from Earth. In Figure 9-5, "phase B" represents a crescent phase of the Moon as viewed from Earth during this period of time.

   Likewise, as the Moon continues moving from point C to point E the illuminated portion continues to get larger in Earth's sky as the Moon moves from the first quarter phase to the full phase. During this portion of its cycle the Moon's phase is known as a gibbous phase. From Earth, the Moon appears larger than a half circle but less than a full circle. The illuminated area is still on the right edge of the Moon as viewed from Earth. In Figure 9-5, "phase D" represents a gibbous phase of the Moon as viewed from Earth while moving from point C to point E.

   As the Moon continues moving from point E to point G the illuminated portion of the Moon becomes less than a full circle but larger than a half circle. The Moon's phase is once again a gibbous phase while moving through this part of its orbit. In Earth's sky, it still appears larger than a half circle again and smaller than a full circle. The illuminated area, however, is now on the left edge now rather than the right edge as was seen earlier. In Figure 9-5, "phase F" represents a gibbous phase of the Moon as it appears from Earth while moving between point E and point G.

   The Moon finally completes its cycle around Earth while moving from point G to point A. The illuminated portion of the Moon begins to appear less than a half circle again. Its appearance in Earth's sky is like that of a crescent phase once again. "Phase H," on the right side in Figure 9-5, represents a crescent phase of the Moon as viewed from Earth during this period of time.

   You may have realized by now that at many places in the Moon's orbit the phases of the Moon are either crescent or gibbous. The fact that the Moon has many of these phases might seem to pose a problem in locating where the Moon is in its orbit. However, there is a way to tell precisely where the Moon is located in its orbit with respect to the Sun.

As the Moon goes around Earth and consequently through its phases, you should notice that when the Moon moves between new and full the illuminated portion becomes larger. By the same token, when the Moon moves between full and new its illuminated portion becomes smaller.

When the Moon moves between new and full the phases are referred to as waxing phases, because they appear to get larger. When the Moon moves between full and new the phases are referred to as waning phases, because they appear to get smaller.

Waxing Phases of the Moon

At point B in Figure 9-1, the phase of the Moon is referred to as a waxing crescent. At point D in Figure 9-1, the phase of the Moon is referred to as a waxing gibbous. In Figure 9-5, "phase B" represents the way a waxing crescent Moon appears in Earth's sky, and "phase D" represents how a waxing gibbous Moon appears. During the waxing stage of the Moon's phase cycle, the right-hand edge of the Moon is visible from Earth. Figure 9-12 displays the waxing crescent phase of the Moon in Leo on June 1, 2006, approximately two hours before the apparent sunset as seen from Ball State Observatory.

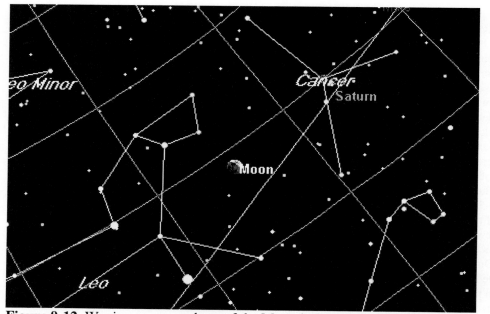

Figure 9-12 Waxing crescent phase of the Moon in Leo on June 1, 2006

Figure 9-13 shows *TheSky's* rendition of what a waxing crescent phase of the Moon looks like when viewed through a telescope on this date. It is a zoomed-in view of the Moon in Figure 9-12. Notice that it resembles "phase B" in Figure 9-5. A waxing crescent Moon appears to rise a few hours after the Sun appears to rise and appears to set a few hours after the Sun. It is visible in Earth's sky in the early evening just after sunset for a few hours.

Figure 9-13 Appearance of a waxing crescent phase of the Moon in *TheSky*

   If you would like to view the sky window displayed in Figure 9-12, then follow the instructions provided in Exercise F.

Exercise F: Observing the Waxing Crescent Phase of the Moon on July 1, 2006

1. Run *TheSky*.

2. Open the multimedia file named WaxingCrescent.sky.

3. Click on the Sun and record its equatorial coordinates here.

    R.A. = _____ ^{H} _____ ^{M} _____ ^{S} Declination = __ _____ $^\circ$ _____ $'$ _____ $''$

4. Find the Moon to display the Object Information window.

5. Center (⊕) the Moon in the sky window.

6. Is the Phase greater than 0% and less than 50%? _____

7. What is the Moon and Sun's angular separation? _____ $^\circ$
  Is it greater than 0° but less than 90°? _____

8. Find the Moon's coordinates and record them here.

    R.A. = _____ ^{H} _____ ^{M} _____ ^{S} Declination = __ _____ $^\circ$ _____ $'$ _____ $''$

9. Compare the Moon's right ascension to that of the Sun. Sun _____ ^{H} Moon _____ ^{H}
  Is the difference between 0^H and 6^H? _____

10. Use the Zoom button to observe the Moon up close as viewing it through a telescope.

11. Close the file and do not save it.

  Figure 9-14 displays the view of a waxing gibbous Moon in Virgo on June 6, 2006, at 18:08, approximately two hours before apparent sunset, as seen from Ball State Observatory.

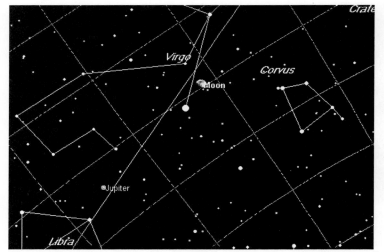
Figure 9-14 Waxing gibbous phase of the Moon in Virgo on June 6, 2006

Figure 9-15 represents *TheSky's* rendition of what a waxing gibbous phase of the Moon looks like when viewed through a telescope on this date. It is a zoomed-in view of the Moon in Figure 9-14. Notice that it is like "phase D" in Figure 9-5. A waxing gibbous Moon appears to rise sometime after midday until just before sunset. It is visible in the eastern sky from the early afternoon until it appears to set sometime after midnight.

Figure 9-15 Appearance of a waxing gibbous phase of the Moon as seen in *TheSky*

If you would like to view the sky window displayed in Figure 9-14, then follow the instructions provided in Exercise G.

Exercise G: Observing a Waxing Gibbous Phase of the Moon on June 6, 2006

1. Run *TheSky*.

2. Open the multimedia file named WaxingGibbous.sky.

3. Click on the Sun and record its coordinates here.

        R.A. = _____^H _____^M _____^S   Declination = __ _____° _____′ _____″

4. Find the Moon to display the Object Information window.

5. Center (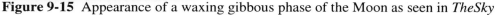) the Moon in the sky window.

6.   Is the Phase greater than 50% and less than 100%? _____

7.   What is the Moon and Sun's angular separation? _____°
     Is it greater than 90° but less than 180°? _____

8.   Find the Moon's coordinates and record them here.

     R.A. = _____^{H} _____^{M} _____^{S} Declination = __ _____° _____' _____"

9.   Compare the Moon's right ascension to that of the Sun. Sun _____^{H} Moon _____^{H}
     Is the difference between 6^H and 12^H? _____

10. Use the Zoom button to observe the Moon up close as viewing it in a telescope.

11. Close the file and do not save it.

Waning Phases of the Moon

At point F in Figure 9-1, the phase of the Moon is referred to as a waning gibbous phase. At point H in Figure 9-1, the phase of the Moon is referred to as a waning crescent. In Figure 9-5, "phase F" represents the way a waning gibbous phase of the Moon appears and "phase H" represents how a waning crescent phase of the Moon appears in Earth's sky. During the waning stage of the Moon's phase cycle, the left-hand edge of the Moon is visible from Earth.

Figure 9-16 displays a view of a waning gibbous phase of the Moon in on June 16, 2006, at 12:45 A.M. EST as seen from Ball State Observatory.

Figure 9-16   Waning gibbous phase of the Moon in Capricornus on June 16, 2006

Figure 9-17 represents what a waning gibbous phase of the Moon looks like when viewed through a telescope on this date. It is a zoomed-in view of the Moon in Figure 9-16. Notice that it resembles "phase F" in Figure 9-5. A waning gibbous Moon appears to rise sometime after midnight and is visible in the eastern sky from early morning until it appears to set sometime after midday.

Figure 9-17 Appearance of the waning gibbous phase of the Moon as seen in *TheSky*

If you would like to view the sky window displayed in Figure 9-16, then follow the instructions provided in Exercise H.

Exercise H: Observing the Waning Gibbous Phase of the Moon on June 16, 2006

1. Run *TheSky*.

2. Open the multimedia file named WaningGibbous.sky.

3. Click on the Sun and record its equatorial coordinates here.

    R.A. = _____^{H} _____^{M} _____^{S} Declination = __ _____$^\circ$ _____$'$ _____$''$

4. Find the Moon to display the Object Information window.

5. Center (⊡) the Moon in the sky window.

6. Is the Phase greater than 50% and less than 100%? _____

7. What is the Moon and Sun's angular separation? _____$^\circ$
    Is it greater than 90° but less than 180° _____

8. Find the Moon's equatorial coordinates and record them here.

    R.A. = _____^{H} _____^{M} _____^{S} Declination = __ _____$^\circ$ _____$'$ _____$''$

9. Compare the Moon's right ascension to that of the Sun. Sun _____^{H} Moon _____^{H}
    Is the difference between 12^H and 18^H? _____

10. Use the Zoom button to observe the Moon up close as viewed in a telescope.

11. Close the file and do not save it.

Figure 9-18 shows the waning crescent phase of the Moon in Pisces on June 20, 2006, at 5:30 A.M. EST, just after the apparent sunrise as seen from Ball State Observatory.

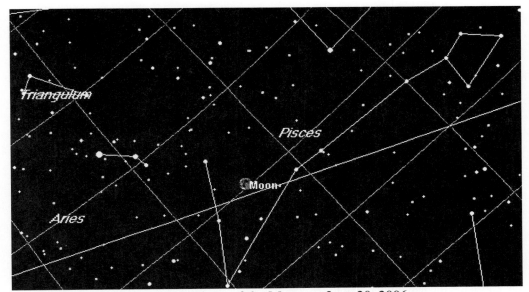

Figure 9-18 Waning crescent phase of the Moon on June 20, 2006

Figure 9-19 represents what the waning crescent phase of the Moon looks like when viewed through a telescope on that morning. It is a zoomed-in view of the Moon in Figure 9-17. Notice that it resembles "phase H" in Figure 9-5. A waning crescent Moon appears to rise a few hours before the Sun and is visible before the Sun appears to rise in the early morning. It appears to set a few hours before the Sun in the late afternoon.

Figure 9-19 Appearance of the waning crescent phase of the Moon as seen in *TheSky*

If you would like to view the sky window displayed in Figure 9-18, then follow the instructions provided in Exercise I.

Exercise I: Observing the Waning Crescent Phase of the Moon on June 20, 2006

1. Run *TheSky*.

2. Open the multimedia file named WaningCrescent.sky.

3. Click on the Sun and record its equatorial coordinates here.

       R.A. = _____^H _____^M _____^S   Declination = __ _____° _____′ _____″

4. Find the Moon to display its Object Information window.

5. Center ( ) the Moon in the sky window.

6. Is the Phase greater than 0% and less than 50%? _____

7. What is the Moon and Sun's angular separation? _____ °
   Is it greater than 0° but less than 90°? _____

8. Find the Moon's equatorial coordinates and record them here.

   R.A. = _____ ^H _____ ^M _____ ^S   Declination = __ _____ ° _____ ' _____ "

9. Compare the Moon's right ascension to that of the Sun. Sun _____ ^H Moon _____ ^H
   Is the difference between 18^H and 24^H? _____

10. Use the Zoom button to observe the Moon up close as in a telescope.

11. Close the file and do not save it.

TheSky displays all lunar phases quite accurately and with amazing clarity. It is easy to determine the phase of the Moon for any date or time or year. The Date/Time feature must first be set to the appropriate date and time in question. You can then find the Moon by accessing the Edit menu on the toolbar at the top of the sky window or right click the right mouse button any where in the sky window. The zoom feature allows to you to observe the phase of the Moon closely and more clearly—a telescopic view so to speak!

Figure 9-20 illustrates the phases of the Moon, which have been rendered in *TheSky*. They are all zoomed frames and look remarkably like the real Moon. The phases in the figure are labeled A through H as they were in Figure 9-5. The lunar phases seen in Figure 9-5 and Figure 9-20 portray one complete phase cycle of the Moon from new Moon to new Moon, a 29½-day cycle.

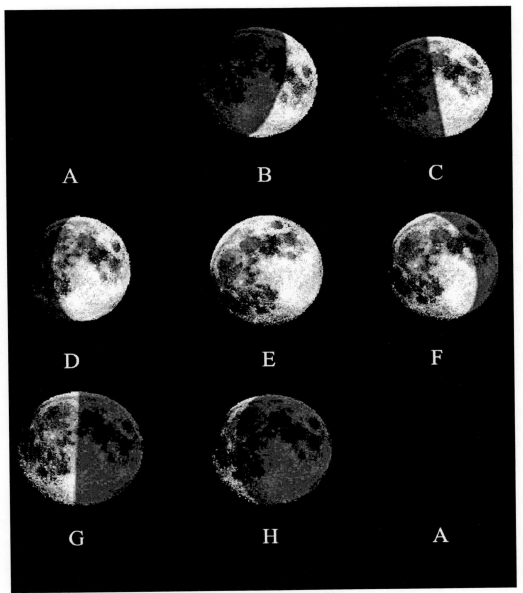

A   B   C

D   E   F

G   H   A

Figure 9-20 Phases of the Moon as displayed in *TheSky*

Exercises J, K, L, M, N, and O are exercises designed to help you use *TheSky* to determine and identify lunar phases.

Exercise J: Determining the Phase of the Moon

1. Run *TheSky* and set as follows:

   Date: October 12, 1492
   Time: 2:30 A.M.
   Daylight savings adjustment option: "Not Observed"
   Location: San Salvador, Island
   Latitude: 24° 24′ 30″ N
   Longitude: 75° 31′ 53″ W

2. Find the Moon and center it.

3. Sketch its appearance (zoom in, if necessary).

4. What is its phase? _____ _____

5. In what constellation is the Moon located? _____

6. What time does it appear to transit the local celestial meridian? _____:_____

7. Close the file and do not save it, or continue with Exercise K.

Exercise K: Determining the Phase of the Moon

1. Run *TheSky* and set as follows:

   Date: July 4, 1776
   Time Zone: +5.0
   Daylight savings adjustment option: "Not Observed"
   Time: 9:30 P.M.
   Location: Philadelphia, Pennsylvania

2. Find the Moon and center it.

3. Sketch its appearance (Zoom in, if necessary).

4. What is its phase? _____ _____

5. In what constellation is the Moon located? _____

6. What time does it appear to transit the local celestial meridian? _____:_____

7. Close the file and do not save it, or continue with Exercise L.

Exercise L: Determining the Phase of the Moon

1. Run *TheSky* and set as follows:

   Date: December 7, 1941
   Time: 6:30 A.M.
   Location: Honolulu, Hawaii

2. Find the Moon and center it.

3. Sketch its appearance.

4. What is its phase? _____ _____

5. In what constellation is the Moon located? _____

6. What time does it appear to transit the local celestial meridian? _____:_____

7. Close the file and do not save it, or continue with Exercise M.

Exercise M: Determining the Phase of the Moon

1. Run *TheSky* and set as follows:

   Date: July 20, 1969
   Time: 8:30 P.M.
   Time Zone: +5.00
     Daylight savings adjustment option: "Not Observed"
     Location: Cape Canaveral, Florida
     Latitude: 28° 29′ 00″ N
     Longitude: 80° 34′ 00″ W

2. Find the Moon and center it.

3. Sketch its appearance.

4. What is its phase? _____ _____

5. In what constellation is the Moon located? _____

6. What time does it appear to transit the local celestial meridian? _____:_____

7. Close the file and do not save it, or continue with Exercise N.

Exercise N: Determining the Phase of the Moon

1. Run *TheSky* and set as follows:

   Date: March 21, 5555
   Time: 2:30 A.M.
   Daylight savings adjustment option: "North America"
   Location: New York City, New York

2. Find the Moon and center it.

3. Sketch its appearance.

4.   What is its phase? _____ _____

5.   In what constellation is it located? _____

6.   What time does it appear to transit the local celestial meridian? _____:_____

7.   Close the file and do not save it, or continue with Exercise O.

Exercise O: Determining the Phase of the Moon

1.   Run *TheSky* and set as follows:

   Date: December 31, 9999
   Time: 2:30 A.M.
   Time Zone: +5.0
   Daylight savings adjustment option: "Not Observed"
   Location: Ball State Observatory.

2.   Find the Moon and center it.

3.   Sketch its appearance.

4.   What is its phase? _____ _____

5.   In what constellation is the Moon located? _____

6.   What time does it appear to transit the local celestial meridian? _____:_____

7.   Close the file and do not save it.

Understanding why the Moon goes through phases and why it appears the way it does in Earth's sky is all a matter of angles. Many people have the notion that the phases of the Moon are caused by shadows. This is simply not true! Phases of the Moon are a direct result of viewing the Moon at different positions in its orbit relative to Earth and the Sun.

Is there an easier way to determine the phases of the Moon? Perhaps the simplest way is to consult a typical calendar. This might take away some of the fun of determining the lunar phase. However, *TheSky* software is equipped with a lunar phase calendar. It displays the phases of the Moon for the current month and year.

In this mode, the previous or next month's phases can easily be viewed for any day of the month for any year. The day on which the Moon's phase is a new Moon, first quarter, full Moon, or last quarter is displayed as well as the time it occurs. The time indicated below that particular phase, is the time at which the Moon reaches that point (new, first quarter, full, last quarter) in its orbit and is expressed in the time at your location.

The Moon Phase Calendar can be printed out, along with the times of the apparent moonrise and moonset. It is a helpful tool provided in the software. Figure 9-21 displays the Moon Phase Calendar in *TheSky* for June 2006. If you would like to use this tool, then follow the instructions provided in the Exercises P and Q.

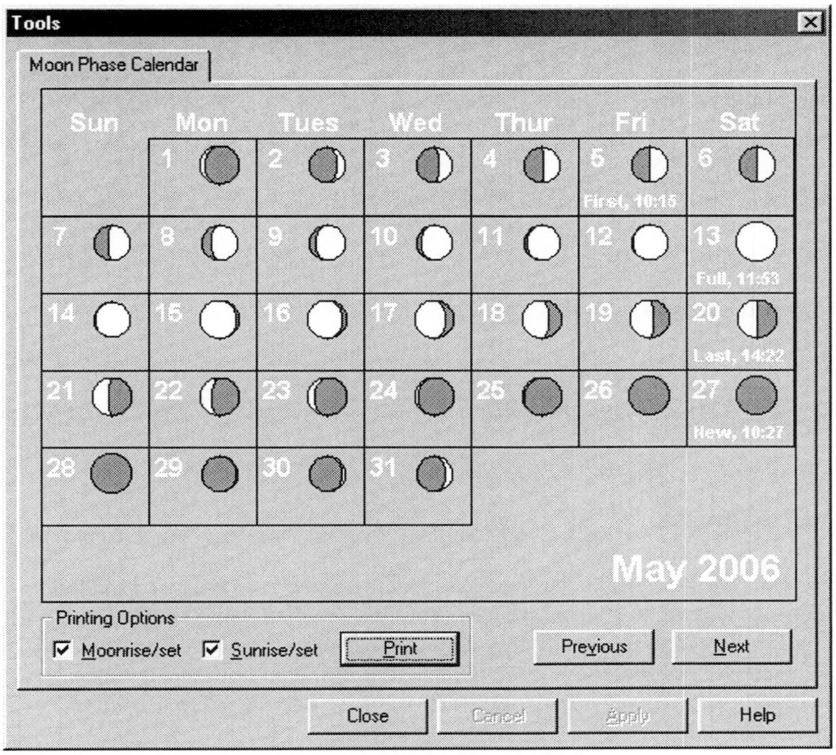

Figure 9-21 Moon Phase Calendar for June 2006

Exercise P: Displaying the Moon Phase Calendar in June 2006

1. Run *TheSky* and set as follows:

   Date: June 1, 2006
   Time: Any value

2. Click *Tools* on the toolbar at the top of the sky window.

3. Click the Moon Phase Calendar.

4. For this month, on what date is the Moon's phase:

   New: _____
   Waxing gibbous: _____
   First quarter: _____
   Waning crescent: _____
   Full: _____
   Last quarter: _____

5. Close the file and do not save it, or continue with Exercise Q.

Exercise Q: Displaying the Moon Phase Calendar for December, 9888

1. Run *TheSky* and set as follows:

   Date: December 1, 9888
   Time: Any value

2. Open the <u>Moon</u> Phase Calendar.

3. On what date is the Moon's phase full in December? _____

4. Close the file and do not save it.

Eclipses of the Sun and Moon

Unlike the phase cycle of the Moon, certain astronomical events *are* directly linked to shadows. These, of course, are eclipses. Earth and Moon are large enough, relatively speaking, to cast shadows in space. With Earth in motion around the Sun and the Moon in motion around Earth, there inevitably arises the possibility that these three objects might line up with one another. When this happens, an eclipse is possible. Eclipses of the Moon are known as lunar eclipses and those of the Sun are called solar eclipses.

Twice a month the Sun, Moon, and Earth align with each other at the time of a new and full Moon. So why don't we observe a lunar and solar eclipse every two weeks? That is, why doesn't Earth or the Moon cast a shadow on the other every time this alignment occurs? Eclipses involve three aspects of Earth, Moon, and Sun. The first and perhaps the most obvious thing is that eclipses depend on the phase of the Moon. Eclipses can only occur at a time when the Moon's phase is new or full. However, even when this happens, it does not necessarily guarantee that an eclipse will occur.

The second thing depends on the geometry of Earth, Moon, and Sun with respect to each other. The Earth, Moon, and Sun must be nearly aligned with one another in space before an eclipse can occur. Every two weeks Earth, Moon, and Sun are in alignment with each other. This occurs when the Moon is at the point in its orbit when it is between Earth and the Sun (new) and again when the Moon is 180° from the Sun (full). Eclipses do not always happen just because Earth, Moon, and Sun are aligned with each other either. The reason is because the Moon's orbit is tilted with respect to Earth's orbit around the Sun.

The last condition has to do with the Moon's orbit itself. Because the Moon's orbit is tilted a little more than 5°, this means that during half of the lunar phase cycle the Moon is above Earth's orbit and during the other half of its cycle the Moon is below Earth's orbit. The two places where the Moon crosses Earth's orbit in the plane of the ecliptic are known as nodes. One node is located where the Moon moves from below Earth's orbit to above it. This point is defined as the *ascending node*. The other point is located where the Moon moves from above Earth's orbit to below it. This point is defined as the *descending node*. If you draw a line between these two nodes across the Moon's orbit connecting them, this line is defined as the *line of nodes*.

Figure 9-22, illustrates the Moon's orbit around Earth and Earth's orbit around the Sun. The line of nodes is depicted as the dotted line. Earth's shadow is also shown in this figure. To have a lunar or solar eclipse occur, *three conditions* must be fulfilled simultaneously. That is, all three must occur at the same time!

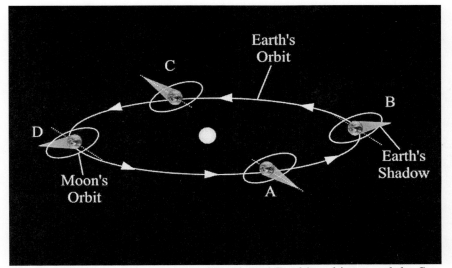

Figure 9-22 Moon's orbit around Earth and Earth's orbit around the Sun

    The first condition is that the Moon's phase *must be* new or full! The second condition is that the Moon *must be* crossing Earth's orbit. In other words, the Moon must be *at* or *near* a node! Finally, the line of nodes *must be* aligned on the Sun. When these three conditions are satisfied at the same instant, and then it is most probable some type of an eclipse will occur. In Figure 9-22 the most likely positions for an eclipse to occur are at positions A and C.

Lunar Eclipses

As was mentioned earlier, Earth and Moon both cast shadows in space. These shadows are, of course, three-dimensional shadows. Figure 9-23 depicts sunlight shining from the left and Earth's shadow being cast into space toward the right. Sunlight passes entirely around Earth and produces two shadows in the direction opposite the Sun. Because light from the left and right edges of the Sun pass around the left and right edges of Earth, respectively, this produces a very dark shadow in space. Light from the left edge of the Sun passes around the right edge of Earth and light from the right edge of the Sun passes around the left edge of Earth, respectively, produces a lighter shadow in space. This geometry is depicted in Figure 9-23. This phenomenon produces one shadow that is quite dark and another one that is much lighter. The darker shadow of Earth, labeled "*A*" in Figure 9-23, is defined as the *umbra*. The lighter shadow of Earth, labeled "*B*" in Figure 9-23, is defined as the *penumbra*. If the Moon passes through either of these shadows, then people on Earth lunar eclipse.

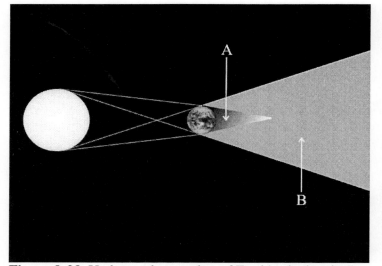

Figure 9-23 Umbra and penumbra of Earth's shadow in Space

Types of Lunar Eclipses

The type of lunar eclipse observed from Earth depends on the part of Earth's shadow through which the Moon moves. If the Moon moves entirely through Earth's umbra, (A) then a *total eclipse* of the Moon is observed. If the Moon just misses the umbra and moves through the part of the umbra and the penumbra, then a *partial eclipse* of the Moon is observed. If the Moon misses the umbral shadow completely and moves only through the penumbra (B) then a *penumbral eclipse* is observed.

Of all these types of lunar eclipses, the most striking is a total eclipse. The Moon is completely immersed in Earth's umbral shadow during a total eclipse. The Moon turns a deep red color at mid-eclipse. The reason for this is that only the red light from the Sun makes its way through Earth's atmosphere and falls on the Moon's surface. Unfortunately, *TheSky* does not render color images of total eclipses. It does, however, display the location of Earth's umbra and penumbra in the sky. Figure 9-24 is an image of the totally eclipsed Moon taken on July 6, 1982. The photo was taken at mid-eclipse. The authors gratefully acknowledge the source of the eclipse pictures as a courtesy of Dr. Dale Ireland's Web site: www.drdale.com and they are used here with his permission.

Figure 9-24 Total lunar eclipse of July 6, 1982
(Photo courtesy of Dr. Dale Ireland)

Figure 9-25 displays the March 24, 1997, partial eclipse of the Moon. The photo was taken at mid-eclipse.

Figure 9-25 Partial lunar eclipse of March 24,1997
(Photo courtesy of Dr. Dale Ireland)

Penumbral eclipses of the Moon are not very exciting to watch. The brightness of the Moon changes so little that the change is virtually undetectable from Earth, penumbral eclipses are rarely of interest to anyone. If you did not *know* the Moon was passing through Earth's penumbra ahead of time, you would be hard pressed to determine whether or not an eclipse was taking place.

Figure 9-26 displays the penumbral eclipse that took place on November 18, 1994. As you can see in the photograph, it is very difficult to determine whether or not an eclipse is occurring. The photo was taken at mid-eclipse.

Figure 9-26 Penumbral lunar eclipse of November 18, 1994
(Photo courtesy of Dr. Dale Ireland)

Solar Eclipses

Figure 9-27 depicts the shadow of the Moon being cast on Earth from space. It shows the path this shadow makes on Earth's surface. Sunlight once again is shining from the left in this figure. As sunlight passes entirely around the Moon, it produces two shadows in the direction opposite the Sun. Like Earth's shadow, the Moon has a darker shadow and a lighter one. The Moon's darker shadow, displayed in the figure, is called the umbra while its lighter shadow is called the penumbra. If observers standing on Earth happen to have the Moon's shadow pass over them, they will witness a solar eclipse.

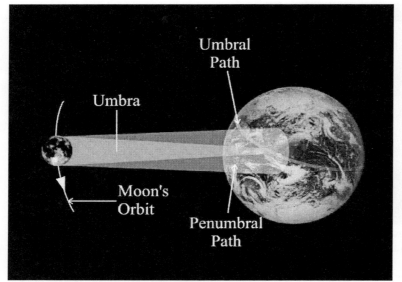

Figure 9-27 – The Umbra of the Moon's Shadow.

Types of Solar Eclipses

Like lunar eclipses, the type of solar eclipse observed from Earth depends on which part of the Moon's shadow passes over the observer. If an observer's location lies *completely within* the umbral path as shown in Figure 9-27, then anywhere along that path a total solar eclipse is observed. Figure 9-28 displays the total solar eclipse that took place on February 26, 1998.

Figure 9-28 Total solar eclipse of February 26, 1998
(Photo courtesy of Dr. Dale Ireland)

Observing total solar eclipses is quite interesting from a scientific point of view. With the bright visible portion of the Sun (known as the *photosphere*) eclipsed, astronomers can study the outer part of the Sun's atmosphere called the *corona*.

If an observer's location is just *outside* the umbral path, but *anywhere within* the penumbral path, then partial eclipses (of varying degrees) may be seen from many locations on Earth. Scientifically, there is not much interest in partial solar eclipses because the bright photosphere of the Sun is only partially blocked by the Moon. Figure 9-29 is an image of the partial solar eclipse taken on February 26, 1998. The picture was taken at the time of mid-eclipse.

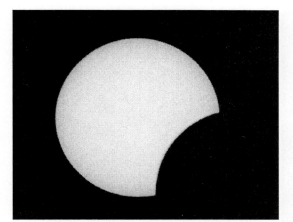

Figure 9-29 Partial solar eclipse of February 26, 1998
(Photo courtesy of Dr. Dale Ireland)

In some instances the Moon's umbral shadow does not completely reach Earth's surface. When this situation arises, the last type of solar eclipse occurs. As the Moon's umbra makes its way toward Earth, it sometimes converges several miles above Earth's surface. Eventually it diverges, however, and finally reaches Earth's surface. An observer standing on Earth looking toward the Sun and Moon in this situation observes the Moon's angular size to be smaller than that of the Sun.

As the Moon moves between Earth and the Sun, from the Earth's surface an observer sees most of the Moon blocking the sunlight except for a small ring of light that encircles it. This small ring is called an annulus. Consequently, this type of solar eclipse is known as an annular eclipse. The geometry of an annular eclipse is illustrated Figure 9-30.

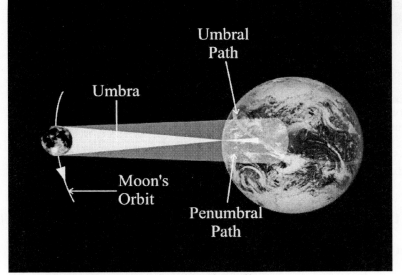

Figure 9-30 Geometry of an annular eclipse

Figure 9-31 displays *TheSky's* rendition of the annular eclipse that took place on May 10, 1994, as viewed from Ball State University campus. It is a zoomed-in view of the Moon. The time of mid-eclipse was 12:07 P.M. EST.

Figure 9-31 *TheSky's* rendition of the annular solar eclipse on May 10, 1994

Figure 9-32 is an actual photograph of the eclipse taken on May 10, 1994 , from Ball State University campus at the time of mid-eclipse. This photograph was taken by Dr. Jordan.

Figure 9-32 Photograph of the annular solar eclipse on May 10, 1994
(Photo Courtesy of Dr. Jordan)

If you would like to view the sky window displayed in Figure 9-31, then follow the instructions provided in Exercise R.

Exercise R: Displaying the Annular Eclipse on May 10, 1994

1. Run *TheSky* and set as follows:

   Location: Ball State Observatory
   Date: May 10, 1994
   Time: 12:07 P.M. EST
   Daylight savings option: "Not Observed"

2. Find the Sun or the Moon and center.

3. Zoom ( ) in on the Sun and Moon.

4. What are the equatorial coordinates of the Sun and Moon?

Sun: R.A.: _____ ^{H} _____ ^{M} _____ ^{S} Declination: __ _____ $^\circ$ _____ ' _____ "

Moon: R.A.: _____ ^{H} _____ ^{M} _____ ^{S} Declination: __ _____ $^\circ$ _____ ' _____ "

5.   Close the file and do not save it.

If an observer's location is outside the penumbral path, then *no* eclipses are observed. *TheSky* does an excellent job of finding and rendering eclipses for any epoch. In fact, the software contains another tool package that finds and illustrates eclipses for any given year. Exercises S, T, U, V and W are designed to help you to use this tool to determine when and where solar or lunar eclipses have and will occur. Note, however, that finding an eclipse with the Eclipse Finder does not guarantee that the eclipse will be visible from a given observer's location.

In the case of solar eclipses, a feature in the Eclipse Finder allows you to observe where the path of totality is located on Earth. Clicking on this feature renders a graphical image of Earth showing where the Moon's path is traced over the area to be eclipsed. It may be necessary to change your location on Earth to view the eclipse. Exercises S, T, U are designed to help you determine several solar eclipses that have and will occur. Exercises V and W are designed to help you determine when several lunar eclipses will occur.

Exercise S: Displaying the Total Solar Eclipse on November 23, 2003

1.   Run *TheSky* and set as follows:

2.   Click *Tools* on the toolbar at the top of the sky window.

3.   Click Eclipse Finder. A window appears like the one displayed in Figure 9-33.

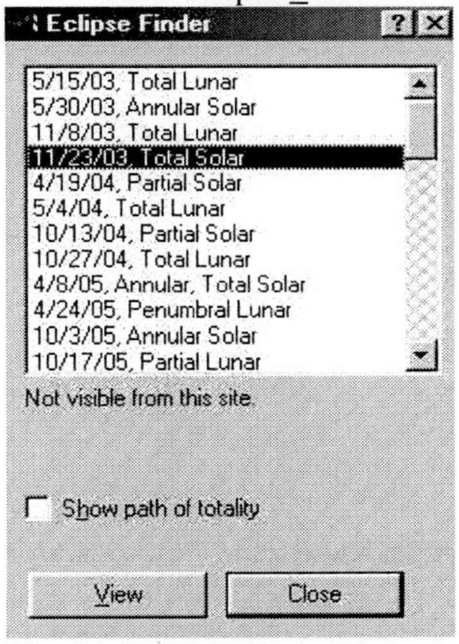

Figure 9-33 Eclipse Finder window

4.   Click the "Total Solar" eclipse on 11/23/03.

5.   Just below the window containing the index of eclipses it states, "Not visible from this site."

6.   Click on the box to "Show path of totality." A window appears like the one displayed in Figure 9-34.

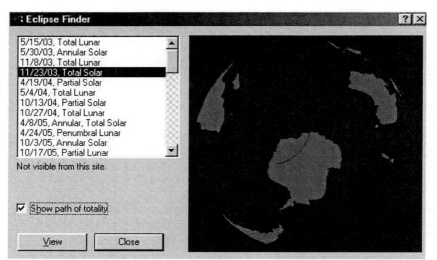

Figure 9-34 Path of totality for the November 23, 2003, solar eclipse

7. The path of totality is across the continent of Antarctica.

8. Click on <u>V</u>iew to watch the eclipse. Notice that the Moon does not move in front of the Sun. The reason for this is that your location is probably not set to Antarctica.

9. If you would like to view this eclipse, then continue with this exercise. Otherwise, close the file and do not save it.

10. Change your latitude and longitude to the following:

    Latitude: 75° S
    Longitude: 86° E

11. Now click the [▶▶] button to watch the eclipse.

12. Close the file and do not save it, or continue with Exercise T.

Exercise T: Displaying the Total Solar Eclipse on August 1, 2008

1. Run *TheSky*.

2. Click <u>T</u>ool on the toolbar at the top of the sky window.

3. Click on Eclipse <u>F</u>inder. A window appears like the one displayed in Figure 9-35.

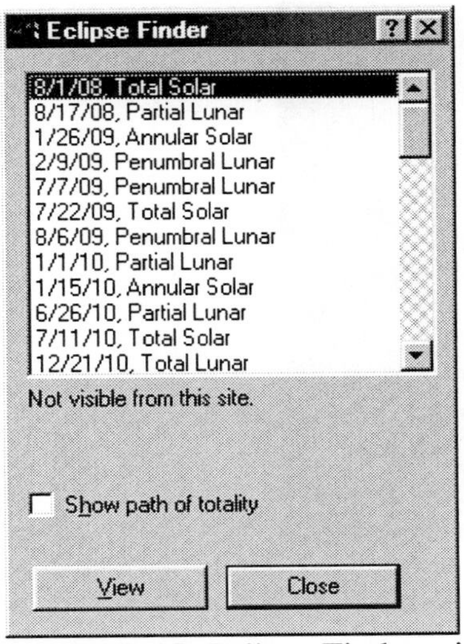

Figure 9-35 Eclipse Finder window

4. Select the "Total Solar" eclipse on August 1, 2008. Notice once again that the eclipse is "Not visible from this site."

5. Check the path of totality.

6. To observe this eclipse, an observer must be in Nanjing, China.

7. Change your location to Nanjing, China as in Figure 9-36.

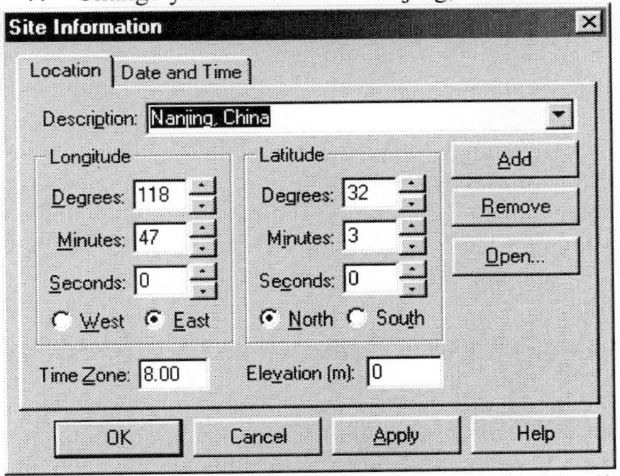

Figure 9-36 Site Information window for Nanjing, China

8. You must open the location file "Cities Outside USA.loc."

9. Click Apply then OK.

10. Click View, and then Close.

11. The screen now appears like the one shown in Figure 9-37.

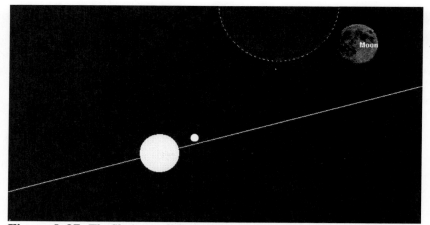

Figure 9-37 *TheSky's* rendition of the total solar eclipse on August 1, 2008, from Nanjing, China

12.   Click 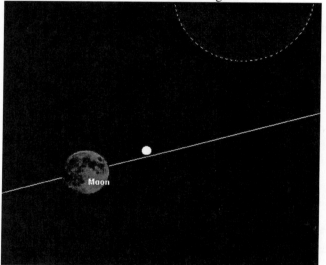 to put the Moon in motion.

13.   Click ◩ to watch the eclipse at 5-minute increments. This slows down the motion of the Moon as is illustrated in Figure 9-38.

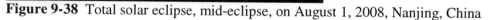

Figure 9-38 Total solar eclipse, mid-eclipse, on August 1, 2008, Nanjing, China

14.   Close the file and do not save it, or continue with Exercise U.

Exercise U: Displaying the Partial Solar Eclipse on December 25, 2000

1.   Run *TheSky* and set as follows:

   Date: Christmas Day, 2000
   Time: 10:45 A.M. EST
   Location: Ball State Observatory

2.   Open the "Eclipse Finder" tool.

3.   Click the "Partial Solar eclipse on 12/25/00, as is shown in Figure 9-39.

4. Click View.

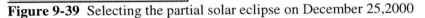

Figure 9-39 Selecting the partial solar eclipse on December 25, 2000

5. The Moon is just right (west) and above (north) of the Sun.

6. Click the [▶▶] button to observe the eclipse. The Moon eclipses the northern half of the Sun as shown in Figure 9-40. Notice too that the Sun and Moon are in the constellation of Sagittarius.

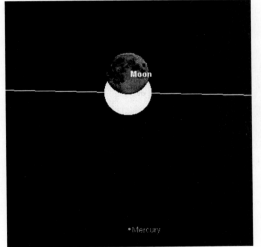

Figure 9-40 Mid-eclipse of the partial solar eclipse on December 25, 2000

7. Close the file and do not save it, or continue with Exercise V.

Exercise V: Observing a Total Lunar Eclipse in 2007

1. Run *TheSky* and set as follows:

   Date: January 1, 2007
   Time: 8:00 A.M.
   Daylight savings option: "North America"
   Location: Oakland, CA.

2. Click Tools on the toolbar at the top of the sky window.

3. Click on the Eclipse Finder and find the next total lunar eclipse.

4. Click the "Total Lunar" eclipse on August 28, 2007, then click View and Close.

5. Now click on Orientation on the Standard toolbar, then Zoom To, then Naked Eye 100°.

6. Now reset *TheSky* as follows:

   Date: August 28, 2007
   Time: 3:45 A.M. PDT

7. The time set is just about mid-eclipse in Oakland, CA. Notice that the Moon is completely immersed in earth's umbral shadow. The eclipse has been rendered in *TheSky* and is displayed in Figure 9-41.

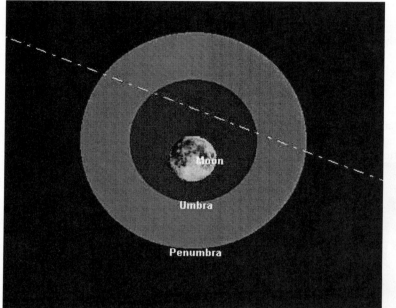

Figure 9-41 Mid-eclipse of the total lunar eclipse on August 28, 2007

8. Changing an observer's location on Earth allows you to view the effects of latitude and longitude on the eclipse. Now change the location to Phoenix, Arizona.

9. Find the Moon. Does anything appear to change from this location? _____

10. Was there any difference in the amount of the Moon eclipsed? _____

11. Phoenix does not observe Daylight Savings Time so change this option to "Not observed."

12. When you change the time setting option, does this change anything for the eclipse? After making the correction for time, does it appear to be the same? _____

13. Now change the location to Chicago, Illinois. Be sure to reset the Daylight savings adjustment option to North America.

14. Does anything appear to change from this location? _____

15. Was there any difference in the amount of the Moon eclipsed? _____.

16. What time does mid-eclipse occur in Chicago, Illinois? _____:_____

17. Finally change the location to New York City. Be sure to *keep* the Daylight savings savings option set to "North America."

18. Does anything appear to change from this location? _____

19. Was there any difference in the amount of the Moon eclipsed? _____

20. What time does mid-eclipse occur in New York City? _____:_____

21. What conclusions might you make about the time at which mid-eclipse occurs for observers at the four different locations? _____

_____

22. Close the file and do not save it, or continue with Exercise W.

Exercise W: Displaying a Partial Lunar Eclipse in 2010

1. Run *TheSky* and set as follows:

   Date: January 1, 2010

2. Click Tools on the toolbar at the top of the sky window.

3. Click on the Eclipse Finder.

4. Click the "Partial Lunar" eclipse on June 26, 2010.

5. Click View and Close.

6. Click the [▶▶] button and watch.

7. Notice that the Moon moves through both the umbra and penumbra of the earth's shadow as illustrated in Figure 9-42.

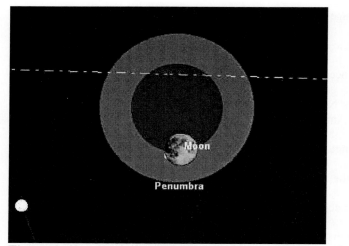

Figure 9-42 Mid-eclipse of the partial lunar eclipse on June 26, 2010

8. To find out when this eclipse occurs at a particular location on Earth, you need to know when a full Moon occurs in June 2010. You can find this information in the Moon Phase Calendar.

9. Reset *TheSky* as follows:

Date: June 26, 2010
Time: 8:00 A.M. EST

10. Click Tools on the toolbar at the top of the sky window.

11. Click the Moon Phase Calendar to display it. It is displayed in Figure 9-43 for June 2010.

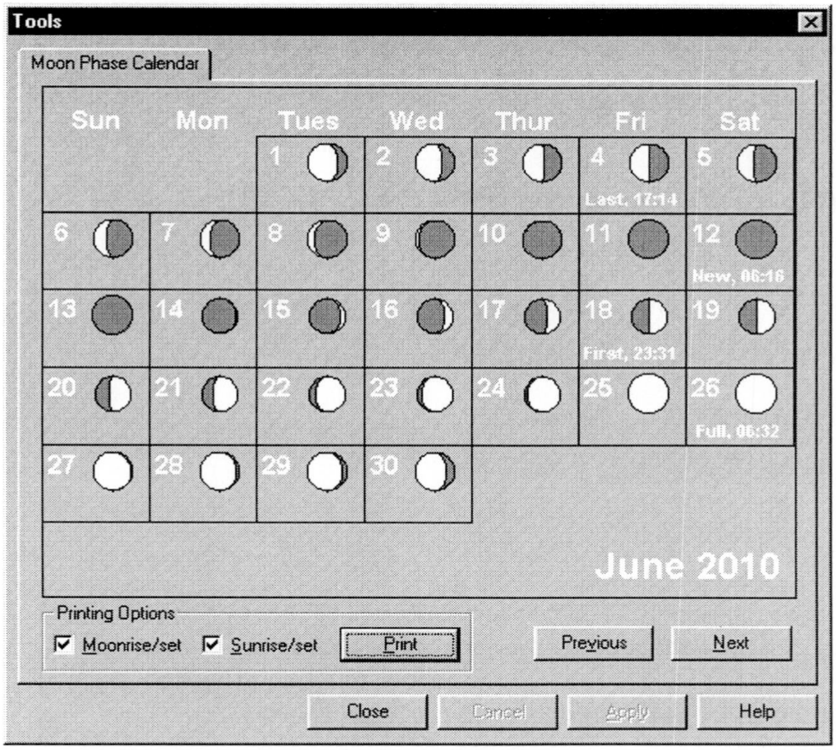

Figure 9-43 Moon Phase Calendar for June 2010

12. Notice that the full Moon occurs on the 26th at 6:32 A.M. EST for an observer at Ball State Observatory. This time is adjusted accordingly for the time zone in which you are located as well as for any corrections for Daylight Savings Time.

13. When does this eclipse occur in Los Angeles, California? _____:_____

14. When does this eclipse occur in Memphis, Tennessee? _____:_____

15. When does this eclipse occur in Bangor, Maine? _____:_____

16. Close the file and do not save it.

Chapter 9

TheSky Review Exercises

Use *TheSky* to complete the following exercises:

TheSky Exercise 1: Determining the Type and Date of an Eclipse

1. What type of eclipse first occurs in the year 2010 C.E.? _____

2. Is it a total, partial, penumbral, or annular eclipse? _____

3. On what date does the eclipse occur? _____

4. Approximately what time does this eclipse occur? _____:_____

5. Is it visible from New York City? _____

6. Is it visible from Edmonton, Canada? _____

7. Is it visible from any part of the United States or Canada on this date? _____
    If so, where? _____

8. If not, where do you have to go on Earth to observe it?

    Latitude = _____°   Longitude = _____°

TheSky Exercise 2: Determining the Date and Time of a Lunar Eclipse

1. On what date does the last lunar eclipse occur in 2010 C.E.? _____

2. Is it a total, partial, or penumbral eclipse? _____

3. On what date does this eclipse occur? _____

4. Approximately what time does this eclipse occur? _____:_____

5. Is it visible from Memphis, Tennessee? _____

6. Is it visible from any part of the United States or Canada? _____

7. If so, where? _____

8. If not, where do you have to go on Earth to observe it?

    Latitude = _____°   Longitude = _____°

TheSky Exercise 3: Determining the Date and Time of a Solar Eclipse

1. On what date did the first solar eclipse take place in the year 1776 C.E.? _____

2. Approximately what time did this eclipse occur? _____:_____

3. Was it a total, partial, or annular eclipse? _____

4. Was it visible from Philadelphia, Pennsylvania? _____

5. Was it visible from any part of the United States or Canada? _____

6. If so, where? _____

7. If not, where did you have to go on Earth to observe it?

    Latitude = _____°   Longitude = _____°

TheSky Exercise 4: Determining the Date and Time of a Lunar Eclipse

1. On what date did the last lunar eclipse occur in the year 1776 C.E.? _____

2. Was it a total, partial, or penumbral eclipse? _____

3. Was it visible from Philadelphia, Pennsylvania? _____

4. Approximately what time did this eclipse occur? _____:_____

5. Was it visible from any part of the United States or Canada? _____

6. If so, where? _____

7. If not, where did you have to go on Earth to observe it?

    Latitude = _____°   Longitude = _____°

TheSky Exercise 5: Determining the Type and Date of an Eclipse

1. On what date does the first eclipse occur in the year 5555 C.E.? _____

2. Is it a lunar or solar eclipse? _____

3. Is it a total, partial, annular, or penumbral eclipse? _____

4. Approximately what time does this eclipse occur? _____:_____

5. Is it visible from the United States or Canada? _____

6. If so, where? _____

7. If not, where do you have to go on Earth to observe it?

    Latitude = _____°   Longitude = _____°

Chapter 9

TheSky Review Questions

Review Questions:

1. The phase of the Moon when it is aligned with the Sun in the sky is called the _____ Moon.

    (a) Full   (b) Last Quarter   (c) First Quarter   (d) New

2. The phase of the Moon when it is located 180° from the Sun is called the _____ Moon.

    (a) Full   (b) Last Quarter   (c) First Quarter   (d) New

3. The phase of the Moon when it is located 90° east of the Sun is called the _____ Moon.

    (a) Full   (b) Last Quarter   (c) First Quarter   (d) New

4. As the Moon moves between a new and a full phase, the illuminated portion of it becomes larger and larger. The phases during this stage of the Moon's phase cycle are known as the_____ phases.

    (a) waving   (b) waning   (c) waxing   (d) wearing   (e) none of these

5. As the Moon moves between a full and a new phase, the illuminated portion of it becomes smaller and smaller. The phases during this stage of the Moon's phase cycle are known as the _____ phases.

    (a) waving   (b) waning   (c) waxing   (d) wearing   (e) none of these

6. The two shadows cast by both Earth and the Moon into space are called the _____ and the _____.

7. What three conditions are necessary for a lunar or solar eclipse to occur?

    (a) _____

    (b) _____

    (c)_____

8. When the Moon moves entirely through the umbra of Earth's shadow, an observer on Earth observes a _____ lunar eclipse.

    (a) partial   (b) annular   (c) penumbral   (d) total   (e)sesquicentennial

9. When the Moon moves partly through the umbra and penumbra of Earth's shadow, an observer on Earth observes a _____ lunar eclipse.

    (a) partial   (b) annular   (c) penumbral   (d) total   (e)sesquicentennial

10. If an observer is standing on Earth and the Moon's umbral shadow passes over them, then that observer observes a _____ solar eclipse.

   (a) partial   (b) annular   (c) penumbral   (d) total

11. If an observer is standing on Earth and the moon's penumbral shadow passes over them, then that observer observes a _____ solar eclipse.

   (a) partial   (b) annular   (c) penumbral   (d) total

12. The type of eclipse that occurs when the moon's umbral shadow doesn't quite reach the ground is called a(n) _____ eclipse of the Sun.

   (a) partial   (b) annular   (c) penumbral   (d) total

The Brightest Stars

Star	Name	Position 2000.0 R.A.	Dec.		Apparent Magnitude	Spectral Type*	Absolute Magnitude	Approximate Distance (LY)
Sol	Sun	-----------	-----------		-26.7	G2V	+4.85	-----------
α CMa A	Sirius	06 45.1	-16	43	-1.44	A0V	+1.44	8.6
α Car	Canopus	06 24.0	-52	42	-0.62	F0Ib	-2.50	95.9
α Boo	Arcturus	14 15.7	+19	11	-0.05	K2IIIp	-0.20	36.7
α Cen A	Rigil Kentaurus	14 39.6	-60	50	-0.01	G2V	+4.37	4.39
α Lyr	Vega	18 36.9	+38	47	+0.03	A0V	+0.60	25.3
α Aur	Capella	05 16.7	+46	00	+0.08	G6III	+0.40	42.2
β Ori A	Rigel	05 14.5	-08	12	+0.18	B8Ia	-8.10	772.9
α CMi A	Procyon	07 39.3	+05	14	+0.40	F5IV-V	+2.70	11.4
α Ori	Betelgeuse	05 55.2	+07	24	+0.45	M2Iab	-7.20	427.5
α Eri	Achernar	01 37.7	-57	14	+0.45	B3Vp	-1.30	143.8
β Cen AB	Hadar	14 03.8	-60	22	+0.61	B1III	-4.40	525.2
α Aql	Altair	19 50.8	+08	52	+0.76	A7IV-V	+2.30	16.8
α Tau A	Aldeberan	04 35.9	+16	31	+0.87	K5III	-0.30	65.1
α Vir	Spica	13 25.2	-11	10	+0.98	B1V	-3.20	262.2
α Sco A	Antares	16 29.4	-26	26	+1.06	M1Ib	-5.20	604.0
α PsA	Fomalhaut	22 57.6	-29	37	+1.17	A3V	+2.00	25.1
β Gem	Pollux	07 45.3	+28	02	+1.15	K0IIIvar	+0.70	33.7
α Cyg	Deneb	20 41.4	+45	17	+1.25	A2Ia	-7.20	3229.3
β Cru	Beta Crucis	12 47.7	-59	41	+1.25	B0.5III	-4.70	352.6
α Leo A	Regulus	10 08.4	+11	58	+1.36	B7V	-0.30	77.5
α Cru A	Acrux	12 26.6	-63	06	+0.77	B0.5IV	-4.20	320.7
ε CMa A	Adhara	06 58.6	-28	58	+1.50	B2II	-4.80	430.9
λ Sco	Shaula	17 33.6	-37	06	+1.62	B1.5IV	-3.50	702.9
γ Ori	Bellatrix	05 25.1	+06	21	+1.64	B2III	-3.90	243.0
β Tau	Elnath	05 26.3	+28	36	+1.65	B7III	-1.50	131.0

*Spectral Types: I = Supergiants, II = Bright Giants, III = Giants, IV = Subgiants, V = Dwarfs

Constellation Names and Their Abbreviations

Latin Name	Possessive Form	Abbreviation	Translation
Andromeda	Andromedae	And	Andromeda[*]
Antlia	Antliae	Ant	Pump
Apus	Apodis	Aps	Bird of Paradise
Aquarius	Aquarii	Aqr	Water Bearer
Aquila	Aquilae	Aql	Eagle
Ara	Arae	Ara	Altar
Aries	Arietis	Ari	Ram
Auriga	Aurigae	Aur	Charioteer
Boötes	Boötis	Boo	Herdsman
Caelum	Caeli	Cae	Chisel
Camelopardalis	Camelopardalis	Cam	Giraffe
Cancer	Cancri	Cnc	Crab
Canes Venatici	Canum Venaticorum	Cvn	Hunting Dogs
Canis Major	Canis Majoris	CMa	Big Dog
Canis Minor	Canis Minoris	CMi	Little Dog
Capricornus	Capricorni	Cap	Goat
Carina	Carinae	Car	Ship's Keel[**]
Cassiopeia	Cassiopeiae	Cas	Cassiopeia[*]
Centaurus	Centauri	Cen	Centaur[*]
Cepheus	Cephei	Cep	Cepheus[*]
Cetus	Ceti	Cet	Whale
Chamaeleon	Chamaeleonis	Cha	Chameleon
Circinus	Circini	Cir	Compass
Columba	Columbae	Col	Dove
Coma Berenices	Comae Berenices	Com	Berenice's Hair[*]
Corona Australis	Coronae Australis	CrA	Southern Crown
Corona Borealis	Coronae Borealis	CrB	Northern Crown
Corvus	Corvi	Crv	Crow
Crater	Crateris	Crt	Cup
Crux	Crucis	Cru	Southern Cross
Cygnus	Cygni	Cvg	Swan
Delphinus	Delphini	Del	Dolphin
Dorado	Doradus	Dor	Swordfish
Draco	Draconis	Dra	Dragon
Equuleus	Equulei	Equ	Little Horse
Eridanus	Eridani	Eri	River Eridanus[*]
Fornax	Fornacis	For	Furnace
Gemini	Geminorum	Gem	Twins
Grus	Gruis	Gru	Crane
Hercules	Herculis	Her	Hercules[*]

Latin Name	Possessive Form	Abbreviation	Translation
Horologium	Horologii	Hor	Clock
Hydra	Hydrae	Hya	Hydra* (Water Monster)
Hydrus	Hydri	Hyi	Sea Serpent
Indus	Indi	Ind	Indian
Lacerta	Lacertae	Lac	Lizard
Leo	Leonis	Leo	Lion
Leo Minor	Leonis Minoris	LMi	Little Lion
Lepus	Leporis	Lep	Hare
Libra	Librae	Lib	Scales
Lupus	Lupi	Lup	Wolf
Lynx	Lyncis	Lyn	Lynx
Lyra	Lyrae	Lyr	Lyre (Harp)
Mensa	Mensae	Men	Table (Mountain)
Microscopium	Microscopii	Mic	Microscope
Monoceros	Monocerotis	Mon	Unicorn
Musca	Muscae	Mus	Fly
Norma	Normae	Nor	Carpenter's Square
Octans	Octantis	Oct	Octant
Ophiuchus	Ophiuchi	Oph	Ophiuchus* (Serpent Bearer)
Orion	Orionis	Ori	Orion* (The Hunter)
Pavo	Pavonis	Pav	Peacock
Pegasus	Pegasi	Peg	Pegasus* (Winged Horse)
Perseus	Persei	Per	Perseus
Phoenix	Phoenicis	Phe	Phoenix
Pictor	Pictoris	Pic	Easel
Pisces	Piscium	Psc	Fishes
Piscis Austrinus	Piscis Austrini	PsA	Southern Fishes
Puppis	Puppis	Pup	Ship's Stern**
Pyxis	Pyxidis	Pyx	Ship's Compass**
Reticulum	Reticuli	Ret	Net
Sagitta	Sagittae	Sge	Arrow
Sagittarius	Sagittarii	Sgr	Archer
Scorpius	Scorpii	Sco	Scorpion
Sculptor	Sculptoris	Scl	Sculptor
Scutum	Scuti	Sct	Shield
Serpens	Serpentis	Ser	Serpent
Sextans	Sextantis	Sex	Sextant
Taurus	Tauri	Tau	Bull
Telescopium	Telescopii	Tel	Telescope
Triangulum	Trianguli	Tri	Triangle

Latin Name	Possessive Form	Abbreviation	Translation
Triangulum Australe	Trianguli Australis	TrA	Southern Triangle
Tucana	Tucanae	Tuc	Toucan
Ursa Major	Ursae Majoris	UMa	Big Bear
Ursa Minor	Ursae Minoris	UMi	Little Bear
Vela	Velorum	Vel	Ship's Sails**
Virgo	Virginis	Vir	Virgin
Volans	Volantis	Vol	Flying Fish
Vulpecula	Vulpeculae	Vul	Little Fox

*Proper Names.
**Use to form the constellation Argo Navis, the Argonauts' Ship

Appendix C

The Greek Alphabet

Letter	Upper case	Lower Case		Letter	Upper Case	Lower Case
Alpha	A	α		Nu	N	ν
Beta	B	β		Xi	Ξ	ξ
Gamma	Γ	γ		Omicron	O	o
Delta	Δ	δ		Pi	Π	π
Epsilon	E	ε		Rho	P	ρ
Zeta	Z	ζ		Sigma	Σ	σ
Eta	H	η		Tau	T	τ
Theta	Θ	θ		Upsilon	Y	υ
Iota	I	ι		Phi	Φ	ϕ
Kappa	K	κ		Chi	X	χ
Lambda	Λ	λ		Psi	Ψ	ψ
Mu	M	μ		Omega	Ω	ω

The Messier Catalogue

M#	NGC#	R. A. (2000)	Dec. (2000)	m_v	Description
1	1952	05 34.5	+22 01	11.3	Crab Nebula in Taurus
2	7089	21 33.5	-00 49	6.5	Globular Cluster in Aquarius
3	5272	13 42.2	+28 23	6.4	Globular Cluster in Canes Venatici
4	6121	16 23.6	-26 32	5.9	Globular Cluster in Scorpius
5	5904	15 18.6	+02 05	5.8	Globular Cluster in Serpens
6	6405	17 40.1	-32 13	5.2	Open Cluster in Scorpius
7	6475	17 53.9	-34 49	3.3	Open Cluster in Scorpius
8	6523	18 03.8	-24 23	5.8	Lagoon Nebula in Sagittarius
9	6333	17 19.2	-18 31	7.9	Globular Cluster in Ophiuchus
10	6254	16 57.1	-04 06	6.6	Globular Cluster in Ophiuchus
11	6705	18 51.1	-06 16	5.8	Open Cluster in Scutum
12	6218	16 47.2	-01 57	6.6	Globular Cluster in Ophiuchus
13	6205	16 41.7	+36 28	5.9	Globular Cluster in Hercules
14	6402	17 37.6	-03 15	7.6	Globular Cluster in Ophiuchus
15	7078	21 30.0	+12 10	6.4	Globular Cluster in Pegasus
16	6611	18 18.8	-13 47	6.0	Open Cluster and Nebula in Serpens
17	6618	18 20.8	-16 11	7.0	Omega Nebula in Sagittarius
18	6613	18 19.9	-17 08	6.9	Open Cluster in Sagittarius
19	6273	17 02.6	-26 16	7.2	Globular Cluster in Ophiuchus
20	6514	18 02.6	-23 02	8.5	Trifid Nebula in Sagittarius
21	6531	18 04.6	-22 30	5.9	Open Cluster in Sagittarius
22	6656	18 36.4	-23 54	5.1	Globular Cluster in Sagittarius
23	6494	17 56.8	-19 01	5.5	Open Cluster in Sagittarius
24	6603	18 16.9	-18 29	4.5	Open Cluster in Sagittarius
25	IC4725	18 31.6	-19 15	4.6	Open Cluster in Sagittarius
26	6694	18 45.2	-09 24	8.0	Open Cluster in Scutum
27	6853	19 59.6	+22 43	8.1	Planetary (Dumbbell Nebula)
28	6626	18 24.5	-24 52	6.9	Globular Cluster in Sagittarius
29	6913	20 23.9	+38 32	6.6	Open Cluster in Cygnus
30	7099	21 40.4	-23 11	7.5	Globular Cluster in Capricornus
31	224	00 42.7	+41 16	3.4	Galaxy in Andromeda (Sb)
32	221	00 42.7	+40 52	8.2	Elliptical Galaxy (Companion M31)
33	598	01 33.9	+30 39	5.7	Spiral Galaxy in Triangulum (Sc)
34	1039	02 42.0	+42 47	5.2	Open Cluster in Perseus
35	2168	06 08.9	+24 20	5.1	Open Cluster in Gemini
36	1960	05 36.1	+34 08	6.0	Open Cluster in Auriga
37	2099	05 52.4	+32 33	5.6	Open Cluster in Auriga
38	1912	05 28.7	+35 50	6.4	Open Cluster in Auriga
39	7092	21 32.2	+48 26	4.6	Open Cluster in Cygnus
40	———	12 22.4	+58 05	8.0	Double Star Winnecke 4 Separation of 50"(UMa)

M#	NGC#	R. A. (2000)	Dec. (2000)	m_v	Description
41	2287	06 47.0	-20 44	4.5	Open Cluster in Canis Major
42	1976	05 35.4	-05 27	4.0	Nebula in Orion
43	1982	05 35.6	-05 16	9.0	Smaller Part of the Nebula in Orion
44	2632	08 40.1	+19 59	3.1	Open Cluster in Cancer (Beehive)
45	———	03 47.0	+24 07	1.2	Open Cluster in Taurus (Pleiades)
46	2437	07 41.8	-14 49	6.1	Open Cluster in Puppis
47	2422	07 36.6	-14 30	4.4	Open Cluster in Puppis, West of M46
48	2548	08 13.8	-05 48	5.8	Open Cluster in Hydra
49	4472	12 29.8	+08 00	8.4	Bright Elliptical Galaxy in Virgo
50	2323	07 03.2	-08 20	5.9	Open Cluster in Monoceros
51	5194/95	13 29.9	+47 12	8.1	Whirlpool Galaxy in CnV
52	7654	23 24.2	+61 35	6.9	Open Cluster in Cassiopeia
53	5024	13 12.9	+18 10	7.7	Globular Cluster in Coma Berenices
54	6715	18 55.1	-30 29	7.7	Globular Cluster in Sagittarius
55	6809	19 40.0	-30 58	7.0	Globular Cluster in Sagittarius
56	6779	19 16.6	+30 11	8.2	Globular Cluster in Lyra
57	6720	18 53.6	+33 02	9.0	Planetary Nebula; Ring Nebula in Lyra
58	4579	12 37.7	+11 49	9.8	Bright Barred-Spiral Galaxy in Virgo
59	4621	12 42.0	+11 39	9.8	Bright Elliptical Galaxy in Virgo
60	4649	12 43.7	+11 33	8.8	Bright Elliptical Galaxy in Virgo
61	4303	12 21.9	+04 28	9.7	Spiral Galaxy in Virgo
62	6266	17 01.2	-30 07	6.6	Globular Cluster in Ophiuchus
63	5055	13 15.8	+42 02	8.6	Spiral Galaxy in Canes Venatici
64	4826	12 56.7	+21 41	8.5	Spiral Galaxy in Coma Berenices
65	3623	11 18.9	+13 05	9.3	Spiral Galaxy in Leo
66	3627	11 20.2	+12 59	9.0	Spiral Galaxy in Leo
67	2682	08 50.4	+11 49	6.9	Open Cluster in Canes Venatici
68	4590	12 39.5	-26 45	8.2	Globular Cluster in Hydra
69	6637	18 31.4	-32 21	7.7	Globular Cluster in Sagittarius
70	6681	18 43.2	-32 18	8.1	Globular Cluster in Sagittarius
71	6838	19 53.8	+18 47	8.3	Globular Cluster in Sagitta
72	6981	20 53.5	-12 32	9.4	Globular Cluster in Aquarius
73	6994	20 58.9	-12 38	—	Open Cluster in Aqr (Group of 4 Stars)
74	628	01 36.7	+15 47	9.2	Spiral Galaxy in Pisces
75	6864	20 06.1	-21 55	8.6	Globular Cluster in Sgr (59,000 LY)
76	650/51	01 42.4	+51 34	11.5	Planetary Nebula in Perseus
77	1068	02 42.7	-00 01	8.8	Spiral Galaxy in Cetus
78	2068	05 46.7	+00 03	8.0	Small Reflection Nebula in Orion
79	1904	05 24.5	-24 33	8.0	Globular Cluster in Lepus
80	6093	16 17.0	-22 59	7.2	Globular Cluster in Scorpius

M#	NGC#	R. A. (2000)	Dec. (2000)	m_v	Description
81	3031	09 55.6	+69 04	6.8	Bright Spiral Galaxy in UMa
82	3034	09 55.8	+69 41	8.4	Irregular Galaxy in Uma (Exploding)
83	5236	13 37.0	-29 52	10.1	Spiral Galaxy in Hydra
84	4374	12 25.1	+12 53	9.3	Elliptical Galaxy in Virgo
85	4382	12 25.4	+18 11	9.3	Elliptical Galaxy in Coma Berenices
86	4406	12 26.2	+12 57	9.2	Elliptical Galaxy in Virgo
87	4486	12 30.8	+12 24	8.6	Elliptical (Peculiar) Galaxy in Vir
88	4501	12 32.0	+14 25	9.5	Spiral Galaxy in Coma Berenices
89	4552	12 35.7	+12 33	9.8	Elliptical Galaxy in Virgo
90	4569	12 36.8	+13 10	9.5	Spiral Galaxy in Virgo
91	4548	12 35.4	+14 30	10.2	Barred Spiral Galaxy in Com (M58?)
92	6341	17 17.1	+43 08	6.5	Another Globular Cluster in Hercules
93	2447	07 44.6	-23 52	6.2	Open Cluster in Puppis
94	4736	12 50.9	+41 07	8.1	Spiral (Peculiar) Galaxy in CVn
95	3351	10 44.0	+11 42	9.7	Barred Spiral Galaxy in Leo
96	3368	10 46.8	+11 49	9.2	Spiral Galaxy in Leo close to M95
97	3587	11 14.8	+55 01	11.2	Planetary Nebula in UMa (Owl Nebula)
98	4192	12 13.8	+14 54	10.1	Spiral Galaxy in Coma Berenices
99	4254	12 18.8	+14 25	9.8	Spiral Galaxy in Coma Berenices
100	4321	12 22.9	+15 49	9.4	Spiral Galaxy in Coma Berenices
101	5457	14 03.2	+54 21	7.7	Spiral Galaxy (Face-on) in Ursa Major
102	5866	15 06.5	+55 46	10.0	M102 = M101? Possible Duplication
103	581	01 33.2	+60 42	7.4	Open Cluster in Cassiopeia
104	4594	12 40.0	-11 37	8.3	Spiral Galaxy in Vir (Sombrero Galaxy)
105	3379	10 47.8	+12 35	9.3	Elliptical Galaxy in Leo (Near M95 & M96)
106	4258	12 19.0	+47 18	8.3	Spiral Galaxy in Canes Venatici
107	6171	16 32.5	-13 03	8.1	Globular Cluster in Ophiuchus
108	3556	11 11.5	+55 40	10.0	Spiral Galaxy in Ursa Major
109	3992	11 57.6	+53 23	9.8	Spiral Galaxy in Ursa Major
110	205	00 40.4	+41 41	8.0	Elliptical Galaxy in Andromeda (Companion M31)

Appendix E

Using Menus, Filters, and Preferences in *TheSky*

This appendix is devoted to help you to set up your sky window. It explains how to use the Filters and Preferences as well as the other features in the View selection at the top of the sky window. *TheSky's* databases are complete catalogues. From these catalogues you may select the types of objects to be displayed on your desktop as well as their limiting magnitudes.

The Filters window determines the celestial objects and the object types you want to display in your sky window. Objects you are not interested in displaying can be hidden. This reduces cluttering on your desktop. The objects listed in the Filters window are listed by type, not alphabetically.

Use the Preference window to create new or personal sky windows or charts. Essentially you can customize the sky window or the sky chart to your own personal tastes. You can change font styles, colors, lines, fill colors, patterns and the like using this window. All of the preference files in *TheSky* folder have an extension "svp." Once you have customized your sky window, then you must use the "Save As" option to save the file. Preference files are located in the *TheSky's* folder "…\user\SVP."

In addition to these items, we also comment on the use and function of other menus used in *TheSky*.

The View Menu

By clicking View on the toolbar at the top of the sky window and then on Filters will display a window like the one shown in Figure E-1. You can also right click on your mouse to get the same window. This window contains a menu that lets you configure the sky window. That is, it will let you display objects on your computer screen. Objects may be clicked on or off at any time. If you wish to display everything on your desktop, simply click the All button. This is illustrated in Figure E-1. This perhaps the best place to start. As you scroll down the list of celestial objects, you may click the box or the check marks to either display the object or not. This will temporarily remove the object from your computer screen.

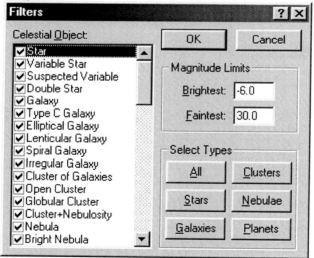

Figure E-1 Filters menu

Selecting a few celestial objects is perhaps the wisest thing to do when you begin using *TheSky*. This will make your desktop less busy and easier to locate stars, planets, and constellations. Typically, you should display the stars, Sun, Moon, planets, and constellation figures at first. Later, as you become more familiar with the program then you can add other items as needed.

On the right half of the Filters window you can set the magnitude limits. The default setting of the software is -6.0 for the brightest and +30.0 for the faintest object. These settings, of course, are not realistic! The brightest star in the sky has a magnitude of -1.5. The Sun, Moon, and planets are displayed regardless of the magnitude settings. The faintest object that can be seen with the naked eye is +6.0 (6th magnitude). Most nonstellar objects are fainter than this but most can be displayed. To change the magnitude settings, you must first highlight the object(s) whose magnitude you wish to change. After highlighting the object(s), then manually type in the magnitudes in the "Brightest" and "Faintest" boxes in the window. For example, you can change the magnitude limits for the stars to represent what is seen when you go outdoors and observe by follow the instructions provided in the first example.

Setting the Magnitude Limits for the Stars in *TheSky*

1. Select View on the toolbar at the top of the sky window, or right click your mouse anywhere in the sky window.

2. Select Filters and highlight the Star in the list, Celestial Object as shown in Figure E-2.

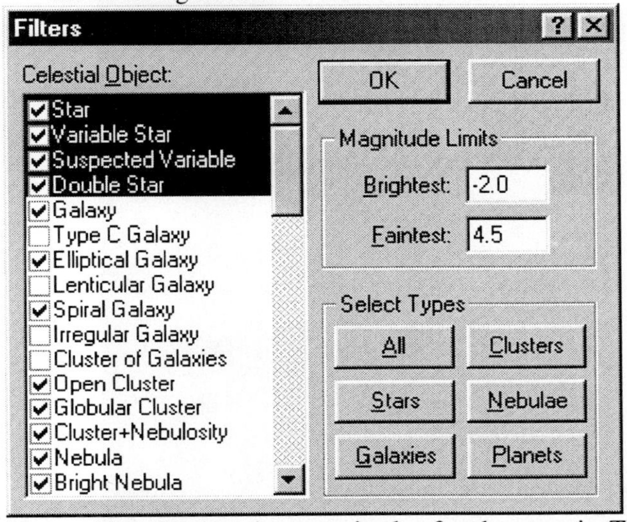

Figure E-2 Setting the magnitudes for the stars in *TheSky*

3. Set the magnitude of the stars in the sky window to −2.0 for the brightest star and +4.5 for the faintest star.

4. Figure E-3 and Figure E-4 displays how the sky window appears before and after, respectively, the changes in magnitudes are made.

Figure E-3 Sky window before setting the magnitude limit

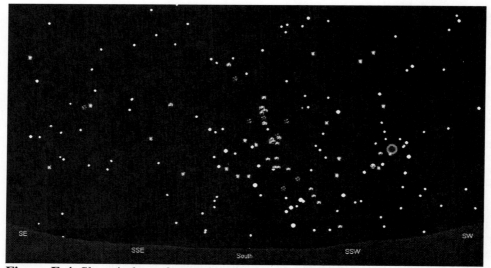

Figure E-4 Sky window after setting the magnitude limits

5. Differences in the two figures are obvious. There are far fewer stars in Figure E-4, as is typically seen in the night sky from a medium-size city.

Figure E-5 shows how my (Dr. Jordan) sky window looks. If needed, I can easily add or remove celestial objects. In fact, the preferences can be set to suit any taste. It is better and less confusing if you configure the sky in a simple fashion at first.

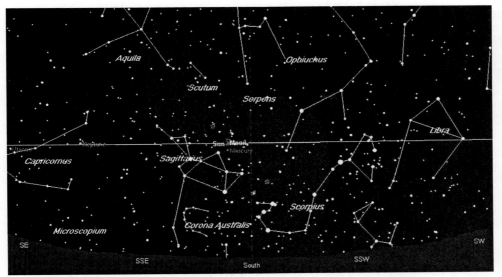

Figure E-5 Dr. Jordan's sky window

We suggest that the beginner display the images in the your sky window. This will enhance your experience using *TheSky*. Displaying images in *TheSky* is accomplished with the F<u>i</u>lter menu. The quickest way for you to make these changes is to right click the mouse anywhere in the sky window. You can also click <u>V</u>iew on the toolbar at the top of the sky window as you have done before. This displays a window that permits you to change the view settings at anytime. After clicking F<u>i</u>lter, you scroll down the index of celestial objects until you find "Image." Once you click the image box in the Filters window, small cameras are displayed in the sky window. As discussed earlier, you can either display these icons or hide them. They are shown in Figure E-6.

Clicking on the camera icons will display images of stellar and nonstellar objects automatically from *TheSky's* database. You can also display images of Solar System objects. You simply click on the object (Sun, Moon, or planet) and then the Multimedia tab in the Object Information window. Accessing stunning images of many of the astronomical objects makes for a pleasant experience in using *TheSky*.

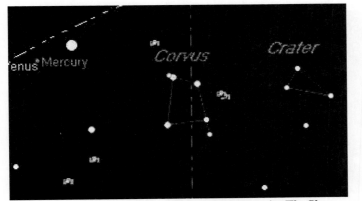

Figure E-6 Camera icons represent images in *TheSky*

It may be impossible to see some of the nonstellar objects with the naked eye, but the images are no less beautiful to look at and appreciate.

Setting the <u>C</u>ommon Names in *TheSky*

Another feature that you will find helpful in the <u>V</u>iew menu, and it must be active, is the <u>C</u>ommon Names option. If you run *TheSky* and your computer screen is filled with stars, but no names or designations are displayed for them, this could seem like a problem. Figure E-7 displays the sky window with no names or designations displayed.

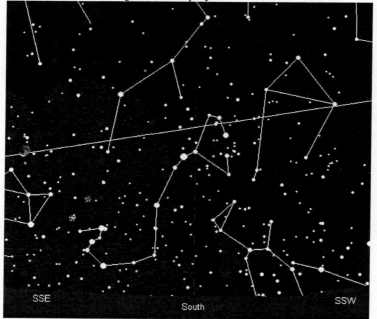

Figure E-7 The sky displayed without the Common Names toggled off

In order to display the common names, you must toggle them on in the <u>V</u>iew menu. The following procedure explains how to toggle on the common names in your sky window.

1. Click <u>V</u>iew on the toolbar at the top of the sky window.

2. Now click L<u>a</u>bels then "<u>C</u>ommon Names."

3. After you click "<u>C</u>ommon Names" the star names, planet names, and constellation names appear in the sky window will appear like the one displayed in Figure E-8.

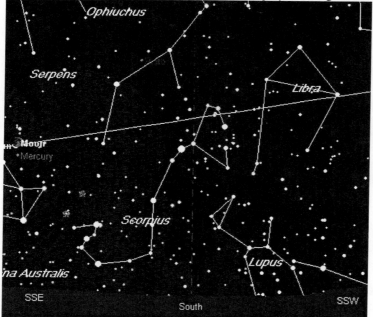

Figure E-8 The sky displayed after toggling on the Common Names

The Labels Setup feature displays the names of astronomical objects in your sky window. You must toggle this option on to display names of stars and constellations. If your computer screen has objects on it, but no names are visible, then you toggle on the Common Names option in the View window.

You select the names of objects that you want to display in your sky window. To accomplish this you must access the Label Setup menu. After clicking on View at the top of your sky window, click on "Labels" then Setup. A menu like the one displayed in Figure E-9 appears.

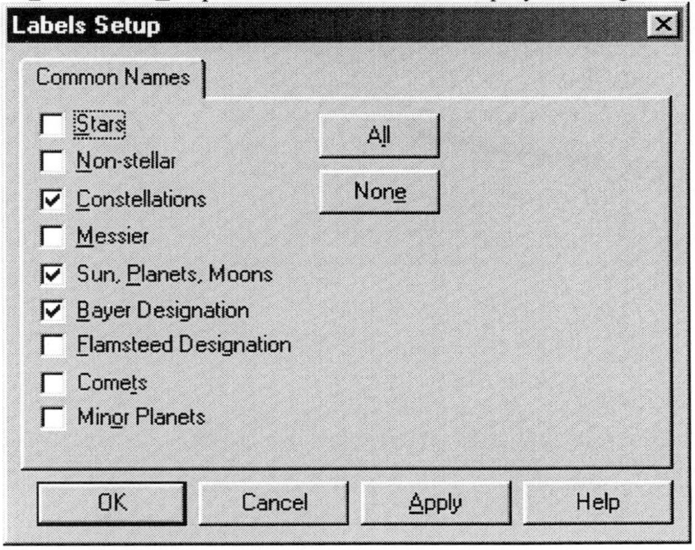

Figure E-9 Labels Setup

This menu window lets you display or hid the names of a variety of things such as stars and constellations. It is also possible for you to display the Messier designations, Bayer Letter designations, and Flamsteed designations with this window. The more names you display, however, the more busy and confusing your sky window becomes.

Displaying Reference Lines

There are other options in the View menu on the toolbar at the top of the sky window. One such option is the Reference Lines. The Reference Lines window lets you display, hide, or change reference lines in your sky window. It is there to simply help you visualize the orientation of and to view portions of the sky you are looking at. You can change the colors of the reference lines in the Preferences window. The following procedure outlines how you can display the reference lines in your sky window.

1. Click View on the toolbar at the top of the sky window, then on "Reference Lines."
   A menu is displayed like the one shown in Figure E-10.

2. When you set up your sky window, you will find it helpful to have several of the General Lines displayed. This might include the Constellation Figures, Ecliptic, and the Milky Way.

3. Other reference lines that are useful are Horizon-based Lines. The Local Meridian is one such line.

4. You will find it convenient to have the horizon displayed in your sky window. You can use the Local Horizon Fill option to set up the horizon in one of two fashions. The first is a Transparent one that allows you to see objects that are below your local horizon. The second one is an Opaque one and resembles what you would actually see when you go outside to observe. You will find it useful to switch from one to the other of these two modes.

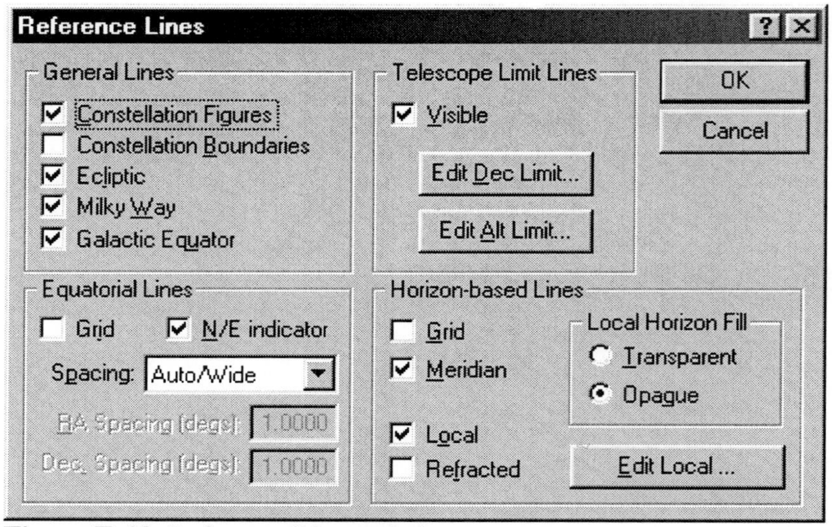

Figure E-10 Reference Lines window

Displaying Mirror Image

The Mirror Image option in the View menu flips the sky window in a left-right fashion. This is the way it would appear in an astronomical telescope.

Displaying the Daytime Sky Mode

The Daytime Sky Mode is used to simulate the apparent rising and setting of the Sun. To use this mode, a time increment must be set beforehand. Usually, a time increment of about 10 minutes is sufficient to observe the apparent short-term motion of the Sun.

Displaying the 3D Solar System Mode

The 3D Solar System Mode displays the Solar System objects in a three-dimensional view. It lets you see where the planets are with respect to the Sun and Earth. More importantly, it lets you see where they are with respect to each other for any given date and time. You can observe there motion with respect to one another.

Displaying the Night Vision Mode

The Night Vision Mode is essential if you decide to use *TheSky* outside or in an observatory. It redraws the toolbars in red to minimize the loss of dark adaptation.

Displaying the Chart Mode

The Chart Mode redraws the sky window with the sky background white and the stars as black dots. It makes the window look more like what would be seen on a star chart or in a book of star atlases. It displays the way the sky window looks when it is printed.

The Preferences Menu

The Preferences in *TheSky* are used to edit the display of objects or the reference lines. As an observational astronomer, knowing that our Sun is a yellow star, I prefer the ecliptic to be represented as a yellow dotted line, rather than the teal color that is the default setting. It is simple to change this feature in the software by following the instructions that follow.

Setting the Ecliptic to a Yellow Dotted Line

1. Click View on the toolbar at the top of the sky window or right click the mouse in the sky window, and then click Preferences. A window is displayed like the one in Figure E-11.

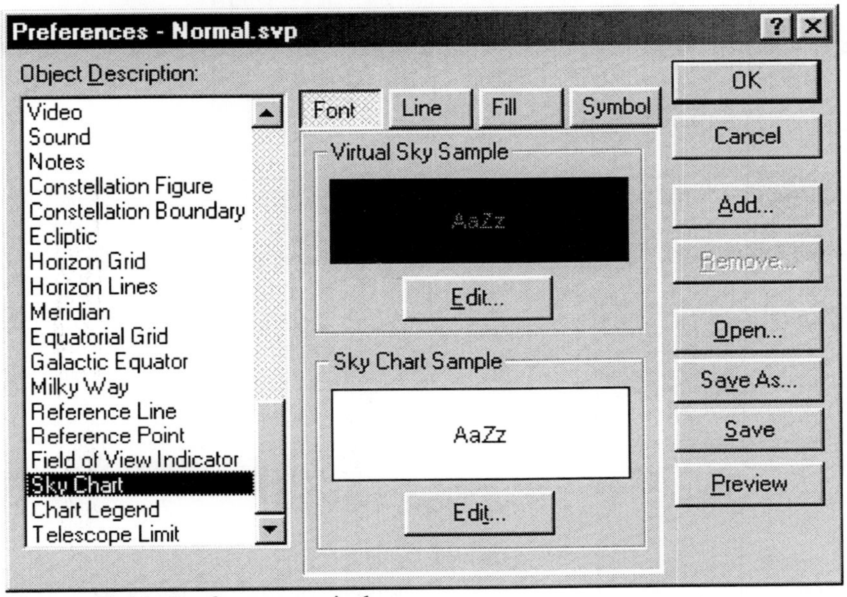

Figure E-11 Preferences window

2. Scroll down the Object Description menu to find the Ecliptic.

3. Click on the ecliptic to highlight it as shown in Figure E-12.

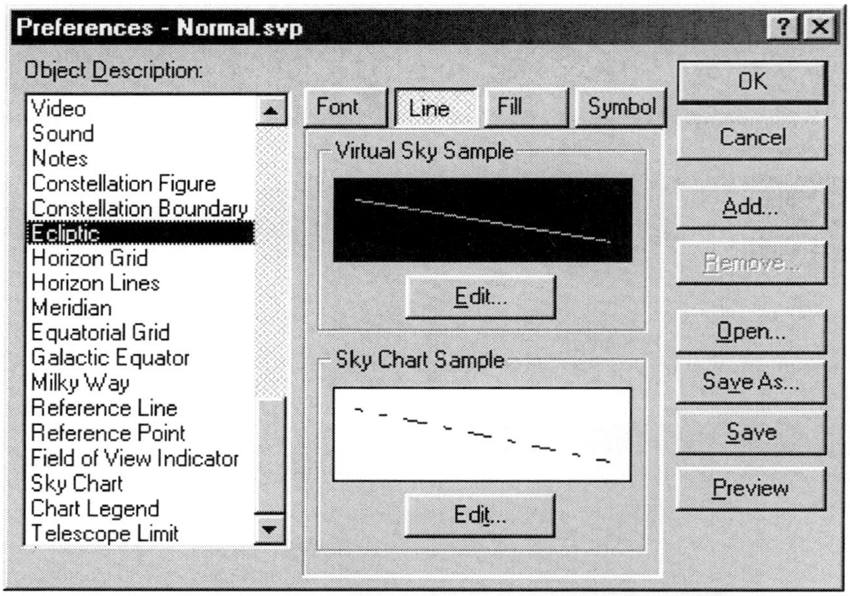

Figure E-12 Changing the color of the ecliptic line in *TheSky*

4. Click the "Line" button then the "Edit" button. A menu like the one displayed in Figure E-13 appears.

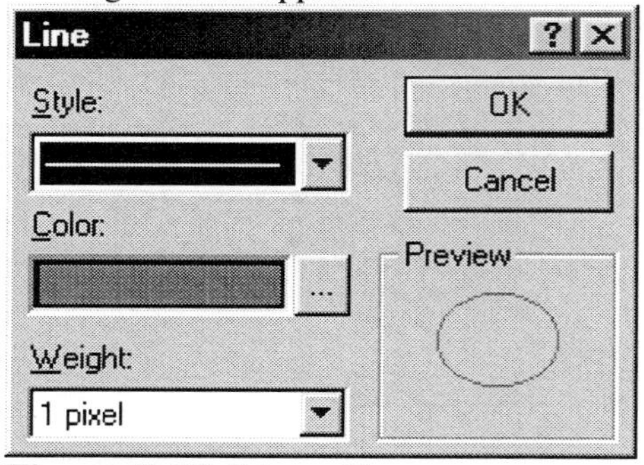

Figure E-13 Line edit menu

5. This menu allows you to change the Style, Color, and Weight (thickness) of a line.

6. To change the line style, click "Style" and change it to a dotted line, and click OK. This returns you back to the Preference menu.

7. Click the color and change it to bright yellow or any other color desired, then click OK. This returns you back to the Preference menu.

8. When the Preference menu is displayed once more, click OK. The ecliptic in your sky window is now displayed as a yellow dotted line.

9. Figure E-14 illustrates the changes that were made to the ecliptic line in your sky window.

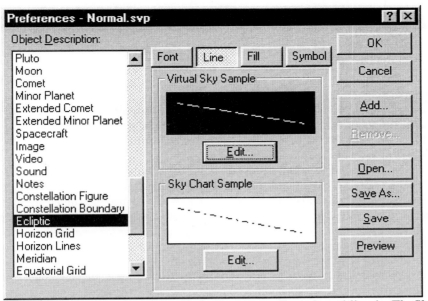

Figure E-14 Changing the ecliptic line to a yellow dotted line in *TheSky*

You may change the appearance of your sky window with this menu at any time. In fact, we recommend that you change the preferences to suit your tastes. If you do not like the change, exit the program without saving the file, and start again.

Displaying the Spectral Colors

The last option in the View menu is the Spectral Colors option. When you select this option the stars are displayed with different hues to simulate their approximate temperature. There are seven major classes of stars according to their temperature. Astronomers classify the stars by their spectral characteristics or absorption lines. The hottest stars are classified as O-stars and appear white in color. B-stars are somewhat cooler and appear blue to blue-white in color. A-stars are cooler still and also appear blue to blue-white in color. F-stars are cooler than A-stars but still appear white to blue-white in color. G-stars have temperatures about that of our Sun and appear yellow in color. K-stars are cooler still than G-stars and appear orange in color. M-stars are the coolest stars and appear reddish orange to red in color.

Figure E-15 lists the seven classes of stars with their approximate temperatures and colors. As you can see, the different classes have temperature ranges. Therefore, subclasses are added to account for these differences and range from 0 to 9 with 0 the hottest and 9 the coolest within each type. The Sun, for example, is classified as a G2 star because of its spectrum (temperature).

Spectral Type:	Temperature:	Color:
O	> 25,000 K	Blue - White
B	11,000 – 25,000 K	Blue – White
A	7,500 – 10,000 K	Blue - White
F	6,000 – 7,500 K	White
G	5,000 – 6,000 K	Yellow - White
K	3,500 – 5,000 K	Orange - Red
M	< 3,500 K	Red

Figure E-15 The spectral classes of stars.

Appendix F

Stellar Magnitudes in *TheSky*

After a casual survey of the nighttime sky, it becomes quite apparent that not all stars appear to have the same brightness. On any clear night, for example, you might be able to see about 3000 to 5000 stars. This, of course, includes the entire sky, both Northern and Southern Hemispheres, and would require observing every night for an entire year! The number of stars seen depends on several things, such as sky conditions, the brightness of the stars, and the observer's vision.

The Greek astronomer Hipparachus developed the first brightness system for stars in the second century before the Common Era (C.E.). He divided stars into six separate groups and assigned numbers to them to represent their apparent brightnesses. To the brightest stars he assigned the number 1, and to the faintest stars, the number 6. Other stars between these two extremes he assigned numbers 2, 3, 4, and 5, respectively. It is a *reversed numbering* system. That is, the *brighter* the star the *smaller* the number, and the *fainter* the star the *larger* the number.

Today we refer to this system as the *magnitude system*. The one we use today differs slightly from the system developed by Hipparchus 22 centuries ago. A magnitude is simply a number assigned to a celestial object that represents the object's apparent brightness.

The brightness is that amount of light energy received per unit area per unit time from a particular object. It is the visible radiant flux coming to us from the object. Some objects only reflect light, such as planets, whereas others generate their own light, such as stars. Regardless of what object(s) you may see, as long as light is received from them, they will have an apparent brightness or magnitude.

The brightness of stars is determined by two factors: the star's luminosity, which is the total amount of electromagnetic energy radiated from its surface into space per unit area per unit time, and its distance from Earth. If you can measure the apparent magnitude of a star and determine its distance from Earth, then you can calculate its total power output or luminosity.

There is a relationship in nature regarding the human senses known as the "Weber-Fechner Law of Psychophysics" (c. 1860). It states that: "*Equal increments of sensations are associated with equal increments of the logarithm of the stimulus.*" This law is said to apply within certain limits to all human sense organs. The eye responds to light in a nonlinear fashion due to the variations in the intensity or brightness of light sources. Simply said, the eye is able to see a tremendous range of light variation.

This system is essentially the basis of the modern magnitude scale adopted by astronomers today. The astronomer N. E. Pogson (c. 1860) suggested that a difference of five (5) magnitudes (actually the difference between a first-magnitude star and a sixth-magnitude star) correspond to a ratio of luminous flux or brightness of 100 to 1. A ratio of brightness corresponding to a step of one magnitude should be the fifth root of 100, which is approximately 2.512.

Because differences of magnitude correspond to ratios of brightness, *differences in magnitude* are *added together* whereas the corresponding *ratios of brightness* are *multiplied together*. If you are acquainted with logarithms, the reason for this rule is clear. This can be easily verified using the values in the Magnitude Scale and multiplying the ratio of brightness for the difference of 3.0 magnitudes by the ratio of brightness for a difference of 5.0 magnitudes.

Magnitude difference of 3.0	→	15.85:1
Magnitude difference of 5.0	→	100.0:1
Their product	→	1585.0

The product of these two ratios of brightness is 1585.0. This is the same value as that of a magnitude difference of 8.0 magnitudes. This is also true for any combination of eight magnitudes such as 3.0 + 5.0, or 4.0 + 4.0, or 6.0 + 2.0, and 7.0 + 1.0 while their corresponding ratios of brightness are multiplied together (try them).

Magnitude Scale

Difference of Magnitude	Ratio of Brightness	Difference in Magnitudes	Ratio of Brightness
0.1	1.096	1.0	2.512
0.2	1.202	2.0	6.310
0.3	1.318	3.0	15.850
0.4	1.445	4.0	39.810
0.5	1.585	5.0	100.000
0.6	1.738	6.0	251.200
0.7	1.905	7.0	631.000
0.8	2.089	8.0	1585.000
0.9	2.291	9.0	3981.000
		10.0	10,000.000
		15.0	1,000,000.000

   The application of the Magnitude Scale may not be apparent (no pun intended) at first glance. The following examples using the Magnitude Scale will make working with stellar magnitudes more clear. Astronomers use these numbers to describe many astronomical objects quantitatively.

   Suppose you observe two stars, let's say Star A and Star B for example. Looking at a star chart or in *TheSky* you determine that the magnitude of Star A is +4.5 and the magnitude of Star B is +2.6. The question arises, which star is brighter?

   If your answer was Star B, this is correct! Remember, the smaller the magnitude number, the brighter the star appears in the night sky. That was easy!

   Now here is a more practical question. *How much* brighter is Star B than Star A? The first thing you must determine is the difference in their magnitudes. If we apply the rule of subtraction (smaller from the larger), we find that +4.5 minus +2.6 is 1.9. The difference in magnitudes of Star B and Star A is 1.9. Once you know the difference in the two stars' magnitudes, the corresponding ratios of brightness can be determined.

   In the Magnitude Scale, no difference of magnitude of 1.9 is listed. What we must do is to look for a combination of two differences in magnitude that add up to 1.9. Adding 1.0 and 0.9 together will give us the difference in magnitudes needed. From these two differences in magnitudes, the corresponding ratios of brightness of 2.512 and 2.291 are found. Multiplying these two ratios together will give us the answer of 5.76. So Star B is 5.76 times brighter than Star A. You could also say that Star A is 5.76 times fainter than Star B. Easy? Of course it is!

   Here is another and more difficult problem to consider. Suppose Star A is 525 times brighter than Star B, what are the two stars' magnitudes? If the difference in magnitudes of Star A and Star B is known and the magnitude of the one of the stars is known, then the other star's magnitude can be determined. However, neither star's magnitude is known in this case. What is known in this example is simply the ratio of brightness between the two stars. The only thing that we know for sure in this example is that the number associated with the magnitude of Star A is smaller than that of Star B.

   We still can make some progress in determining something about the magnitudes of the two stars in this example, however. The first thing we must do is to determine the difference in magnitudes between the two stars. Since the ratio of brightness of the two stars is known (525), we can easily find their difference in magnitudes from this ratio of brightness. Looking at the Magnitude Scale we notice that the brightness ratio of 525 is not listed. We must find the next smallest ratio of brightness to it (namely 251.2). If you divide the ratio of brightness of 525 by 251.2, you will get the ratio of brightness of 2.089. You can verify this by multiplying 2.089 by 251.2, when rounded off equals 525! The product of these two ratios of brightness yields the original ratio of brightness of 525. What we have found in this instance are two numbers in the ratio of brightness column whose product is the given ratio of brightness of the two stars.

   In the difference in magnitude columns, opposite these two multiplicands, are two differences in magnitude when added together correspond to the magnitude difference between Star B and Star A. That is, for a ratio of brightness of 525 the corresponding differences in magnitudes are 6.0 and 0.8. Adding these two numbers together gives the difference in magnitudes between Star A and Star B. Now that the magnitude difference is known, it is possible to find the magnitude of the two stars provided that the magnitude of one of the stars is known.

If the magnitude of Star B is known, then the difference in magnitudes of 6.8 is subtracted from the magnitude of Star B to obtain the magnitude of Star A. The reason, Star A is brighter than Star B. The magnitude of Star A is represented by a smaller number and is obtained from the subtraction of the difference in magnitudes from the magnitude of Star B.

If, on the other hand, the magnitude of Star A is known, then difference in magnitudes 6.8 is added to the magnitude of Star A to obtain the magnitude of Star B. The reason, Star B is fainter than Star A. The magnitude of Star B is represented by a larger number and is obtained from the addition of the difference in magnitudes to the magnitude of Star A.

So far, only the discussion of apparent magnitudes has been given. The *apparent magnitude* of a star is defined as the *brightness of a star as seen from Earth*. The apparent magnitude of a star does not tell us much about any of the physical characteristics of stars themselves.

Remember the two things that influence a star's apparent brightness—the star's luminosity and distance. If we know a star's distance, then the star's total energy output (luminosity) can be determined. A star's luminosity is determined by two unique properties of the star itself, its temperature and size. Stellar sizes and temperatures are determined from radiation laws and spectroscopic analysis. Astronomers determine the luminosities of stars and compare them. To accomplish this, astronomers visualize all the stars at the same distance from Earth and then re-estimate their apparent brightnesses.

The inverse square law of light is used for these determinations. The farther away a light source is from the observer, the fainter it appears by a factor of one over the distance squared. The brightness determined for stars at this so-called standard distance is defined as the star's absolute magnitude. The *absolute magnitude* of a star is defined as the *star's apparent brightness as seen from a distance of 10 parsecs*. A parsec is a distance equal to 3.26 light years or 206,265 astronomical units. It also equals about $3.1 \cdot 10^{13}$ kilometers or about $1.9 \cdot 10^{13}$ miles.

Once you know the absolute magnitude of a star, then calculations like those that were done with apparent magnitudes can be done. However, in this case, the ratios of luminosities are compared instead of ratios of brightness!

In *TheSky* apparent magnitudes for most astronomical objects are provided in the Object Information window. Unless the magnitude of the object is brighter than sixth magnitude (naked-eye limit), it is next to impossible to see without a pair of binoculars or a small telescope. Many of the Messier objects are naked-eye objects, some are not. Unless you have an excellent dark site to make your observations, many Messier objects cannot be seen without the aid of some type of instrument.

TheSky lets you set the magnitude limits on your desktop in the Views menu. Simply click on Views, at the top of the sky window then, Filters. A window such as the one displayed in Figure F-1 appears. You can set the magnitude limits for all of the objects observed in *TheSky* with this menu.

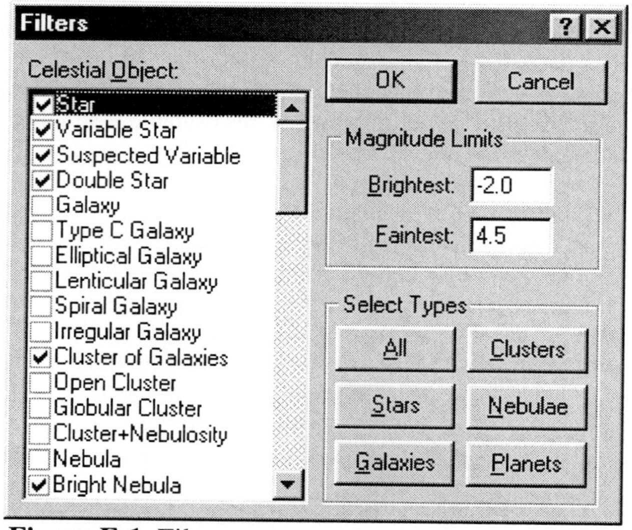

Figure F-1 Filters menu

Appendix F

Review Exercises

Use the Magnitude Scale to answer the following exercises

1. A +4.2 magnitude star is _____ times brighter than a +5.7 magnitude star.
    (Answer: 3.98 times brighter)

2. A −1.7 magnitude star is _____ times _____ than a −2.7 magnitude star.
    (Answer: 2.51 times fainter)

3. A +0.8 magnitude star is _____ times _____ than a +1.3 magnitude star.

4. A 3.7 magnitude star is _____ times _____ than −1.2 magnitude star.

5. Star A is 400 times brighter than star B whose magnitude is +6.3.
    The magnitude of star A is _____ (Answer: −0.2)

6. Star A is 63 times fainter than star B whose magnitude is −0.1.
    The magnitude of star A is _____

7. Venus at its greatest brilliancy has a magnitude of −4.6. Polaris, however, has an apparent magnitude of only +2.0.

    Is Venus brighter or fainter than Polaris? _____
    How many times brighter or fainter? _____

8. What significance does this have as we view Venus and Polaris? _____
    _____

9. How much brighter is the Sun than a full Moon? _____

Review Questions

Answer the following Review Questions about Magnitudes:

1. What does the magnitude of a star measure? _____.

2. What is the apparent magnitude of a star? _____
   _____.

3. What is the absolute magnitude of a star? _____
   _____.

4. Is a star of magnitude of −2 brighter or fainter than a star of magnitude +2?
   _____.

5. How much brighter is a first-magnitude star than a sixth-magnitude star? _____ times.

6. What are the apparent magnitudes of the brightest and faintest stars which can be seen with the naked eye, respectively in the nighttime sky? _____ and _____ magnitude.